THE INNOVATING FIRM

The Innovating Firm

A Behavioral Theory of Corporate R & D

NEIL M. KAY

Foreword by
Christopher Freeman

St. Martin's Press New York

St. Martin's Press, Inc. 175 Fifth Avenue, New York, NY 10010
Printed in Great Britain
First published in the United States of America in 1979

ISBN 0-312-41809-4

Library of Congress Cataloging in Publication Data

Kay, Neil M
The innovating firm.

A revision of the author's thesis, University of Stirling, 1976.
Bibliography: p.
Includes index.
1. Technological Innovations. 2. Research, Industrial. I. Title.
HD45.K37 1979 658.5′7 78-26153
ISBN 0-312-41809-4

To Bill, Elsie and Morris

Contents

Foreword

Two areas of economic theory have recently been undergoing some very deep transformations because of the growing recognition that technical change had received too little attention in previous theoretical formulations. One such area is international trade, where the work of Posner and Hufbauer has led to some major revisions in the theory of comparative advantage and factor costs. Another is the theory of the firm.

For a long time there has been dissatisfaction with the traditional theory of the rational profit-maximising entrepreneur. The discontent has taken many different forms. One very important stream of criticism and research has concentrated on the actual decision-making process within firms. Another has questioned both the centrality and the feasibility of profit maximisation as the main goal of the firm. Still others have pointed to the limitations on 'rationality' in a world where information is very imperfect, and the costs of acquiring information may be high.

From this ferment of ideas, 'behavioural' theories have been developed which attempt in their various ways to provide a more realistic basis for understanding decision-making in the firm. The recent award of a Nobel Prize to Herbert Simon is perhaps one indication that the complexity and importance of this problem is now widely recognised.

However, Simon's work refers very little to R & D decision-making or the specific problems of technical change. A more substantial contribution in this field has been made by Nelson and Winter, whose evolutionary model for the first time provides full scope for Schumpeterian competition, and the search process involved in technical change.

One of the difficulties with the traditional theory of the firm was its tendency to emphasise discrete separate decisions rather than long-term strategies. I therefore particularly welcome Neil Kay's original approach because of its attempt to go beyond the individual decision in an area of central importance for firm behaviour – the allocation of resources to R & D. Like Nelson and Winter he is pioneering a new approach to an old problem and like them I believe he has blazed a trail that many will follow.

CHRISTOPHER FREEMAN

Preface

This book is based on a thesis submitted to the University of Stirling in October 1976 for the degree of Ph.D. The material has been edited to create a hopefully more digestible publication, and some minor changes made in presentation and commentary. Otherwise the content is essentially unchanged.

The book is concerned with the allocation of resources to R & D in the modern corporation. However, equal if not greater importance is attached to the frame of reference developed to study resource allocation in the firm. A general systems framework is developed which, it will be argued, may be a more satisfactory basis for the study of economic behaviour under certain circumstances than conventional approaches based on aggregation.

The thesis itself originated through a strong interest in Cyert and March's behavioural theory of the firm while I was an undergraduate in the Economics Department of Stirling University. As a new department with young staff, the teaching of economics was approached in a fresh and stimulating manner with particular emphasis in microeconomics being laid on the 'Carnegie School'. My first year as a postgraduate of Stirling was spent attempting to apply behavioural theory to problems of innovation. Some of the problems I encountered are incorporated in the discussion of behavioural theory in Chapter 3.

As far as recording my gratitude to the numerous individuals who contributed both directly and indirectly to the development of the analysis, it is difficult to be both fair and comprehensive. J. R. Davies provided both encouragement and useful references when my interest was first stimulated in the economics of innovation as an undergraduate, while C. Sutton, Professor F. R. Bradbury and D. T. Ulph provided useful advice on the early stages of the research. D. Taylor also provided helpful clarification of certain mathematical aspects of the work, while Professor B. J. Loasby provided initial encouragement. Recently, B. Chiplin and R. Young have provided useful criticism of later drafts of the analysis. F. Swenson generously spent some time extending an index used in Chapter 7, without which it is doubtful if the work of that chapter could have been carried out.

I owe a special debt to R. W. Shaw and A. Ulph, both of whom have supervised the thesis. Their comments and criticisms, both encouraging and sceptical, have been of immeasurable assistance in the development and completion of the thesis. Priscilla Baily, Lynda Waters and Pat Lean typed the manuscript, and to them my gratitude for their patience, efficiency and uncanny ability to decipher my handwriting.

December 1977 N. KAY

Acknowledgements

The author and publisher wish to thank the following who have kindly given permission for the use of copyright material:

The American Economic Association and F. M. Scherer for a table from 'Firm size, market structure, opportunity and the output of patented inventions', by F. M. Scherer, published in the *American Economic Review*, 1965; Basil Blackwell & Mott Ltd for a figure from 'Rivalry in industrial research and development: an empirical study, 1973', by H. G. Grabowski and N. D. Baxter, and a table, 'Inter-industry differences' from the *Journal of Industrial Economics*, Vol 21; The Brookings Institution for the material from *Technology, Economic Growth and Public Policy*, by R. R. Nelson, M. J. Peck and E. D. Kalacheck; the *Canadian Journal of Economics* for material from 'Determinants of R & D expenditures in Canadian industry and implications for public policy', ed. L. Waverman; Manchester Business School for an extract from *R & D Budgeting in Swedish Firms*, by N. Kay, published in their journal, Vol 7, 1 October 1976; the National Science Foundation for the figures from *Research and Development in Industry*, 1970 and 1971, in the Surveys of Science Resources Series; Penguin Books Ltd for a table from *The Economics of Industrial Innovations*, 1974, by C. F. Freeman, (Penguin Modern Economic Texts, 1974) © Christopher Freeman 1974; Penguin Books Ltd and George Braziller Inc for a table and a figure from *General System Theory* by Ludwig Von Bertalanffy (Allen Lane The Penguin Press, 1972) © Ludwig Von Bertalanffy, 1968; Prentice-Hall, Inc for a figure from *Corporate Control and Business Behavior*, 1970, by Oliver E. Williamson, © 1970; and the University of Chicago Press for a table from *The Determinants of Industrial Research and Development*, 1968, by H. G. Grabowski.

PART I

1 Introduction

The analysis of economic aspects of technological change is a fairly novel preoccupation of economists. Until recently there has been a conspicuous disregard of this topic by economic theorists, despite its recognised importance in industrial competition and economic growth. As Jewkes et al. (1969) point out, future historians will no doubt find it remarkable that so little systematic analysis was conducted in this area in the first half of the twentieth century. While the post-war period has seen a gradual development of active interest on the part of economists, it still remains a comparatively neglected topic.

The most obvious and possibly most important reason for this neglect is the difficulty involved in adapting and applying conventional economic theory to this area. Innovation is, to a greater or lesser extent, a venture into the unknown as far as its development is concerned. In such circumstances, past experience and quantitative techniques may provide minimal guidance for decision-makers.

It is in this context, and with due acknowledgement of the difficulties faced in model-building in this area, that an approach to the analysis of resource allocation to research and development in the corporation is developed later in the book. In particular, R & D budgeting techniques, possible determinants of the level of R & D expenditure, and the distribution of resources to basic research will be studied in later chapters. It is hoped that the hypotheses tested in this respect will contribute towards an improved understanding of the nature of technological change in the modern corporation.

The book is divided into Part I, consisting of Chapters 1 to 5 inclusive, and Part II, consisting of the last four chapters. Part I is concerned with the development of a framework for the analysis of corporate decision-making, the framework itself being formulated in Chapter 5. The early chapters lay the ground for this development by dealing with three related topics on corporate R & D. Chapter 2 deals with the characteristics of research and development at project level, particularly the relationship between science and technology, and the pervasiveness of uncertainty in R & D work. In this latter respect it provides a basis for criticism of the neoclassical and statistical theories discussed in Chapter 3. However, Chapter 2 also develops the concept of hierarchical arrangement, or ordering, of R & D subsystems which is of use later in analysing the determinants of basic research activity.

Chapter 3 discusses the problems of theory application in conditions of pervasive uncertainty. Three of the approaches discussed – neoclassical economics, decision-making under uncertainty, and behavioural theory – have a common bond in that they were initially developed and applied to problems other than research and development, but each has subsequently been suggested to be applicable to problems of R & D and technological change. A fourth economic approach developed by Penrose (1959) to deal with the growth of firms is also discussed, partly because of its potential application to the area of technological change, but also because it provides useful guidelines for subsequent theory-building, in conjunction with the behavioural theory of the firm.

Chapters 2 and 3 together are intended to demonstrate the difficulties of applying received theory to technological change. Criticism is directed to theory application in this specific area and is not intended to be general criticism of the theories as such. It is in this context that Chapter 4 takes a wider look at corporate decision-making and resource allocation, with special reference to the role of technological change in this framework. The typically hierarchical nature of corporate resource allocation is pointed out, and the role of R & D as a specialised and institutionalised function in the modern corporation is argued. It is suggested that corporations must be regarded as open systems 'which maintain themselves through constant commerce with their environment, i.e. a consistent inflow and outflow of energy through permeable boundaries' (Katz and Kahn, 1966, pp. 18–19.) In this book this is interpreted to mean that decision-makers not only react to the corporate environment, but consciously and autonomously act to shape and mould the environment itself.

This provides the basis for the development of the framework in Chapter 5. The arguments developed in the previous chapter contribute to the concept of the firm as a hierarchically organised open system in which R & D operates as a specialised function.

Part II applies this open system interpretation in the empirical analyses of Chapters 6 to 8 inclusive. Chapter 6 is concerned with accounting for dissimilarities in budgeting conventions adopted by corporations operating under different circumstances in Western Europe and the United States. The evidence of a number of surveys and studies is considered in this chapter, and it is suggested that not only does the open system interpretation reconcile apparently arbitrary differences in budgeting 'style', but also that the systems interpretation developed here provides a rational basis for rule-of-thumb budgeting techniques employed by many large corporations, and frequently described as 'illogical' or 'irrational'.

In Chapter 7, hypotheses based on the open system framework are developed, and regression analysis conducted in an empirical examination of the hypotheses. Specifically, possible determinants of R & D and basic research activity in U.S. industry are investigated. As well as constituting an empirical study of the possible influences on corporate allocations to technological change, it is suggested that the framework developed in Part I provides a sound basis for the empirical hypotheses of this chapter. Not only does it avoid many of the conceptual difficulties of conventional approaches such as neoclassical theory, but it also illustrates how the hierarchic 'top-down' system of resource allocation widely adopted by large corporations may be interpreted as rational behaviour.

Chapter 8 is concerned with intra-industry variation in allocations to technological change activity, and may be regarded as complementing the essentially industry-level orientation of Chapter 7. It discusses the role of rivalry as far as competition in innovative activity is concerned, in particular the propensity of corporations to match or imitate competitor allocations, e.g. in terms of percentage of sales allocated to research and development. Apparently contradictory evidence as to whether or not competitive matching is a prevalent form of industrial behaviour is analysed in this chapter, and a reconciliation is suggested based on the adaptive learning aspect of the systems approach developed earlier.

Part II is concluded with a short summary as to the main conclusions of the analysis and possible implications for future analysis. It is suggested that neoclassical economics does not provide an adequate framework for investigation of technological change, and that the alternative approach developed here and based on concepts developed in general system theory may generate a more satisfactory basis for the study of certain aspects of this problem area. While the difficulties of applying neoclassical economics in this area are generally agreed, it is hoped that the potential usefulness of general system theory is demonstrated through the studies discussed in Chapters 6 to 8 inclusive.

It will be emphasised, however, that the usefulness of the systems approach can only be examined indirectly. No precise, testable hypotheses are provided. As Katz and Kahn (1966) comment: 'In some respects open-system theory is not a theory at all; it does not pretend to the specific sequences of cause and effect, the specific hypotheses and tests of hypotheses which are the basic elements of theory. Open-system theory is rather a framework, a meta-theory' (p. 452).

The approach provides a frame of reference within which lower-level hypotheses capable of empirical testing can be generated. The concepts and interpretations of the systems approach provide a perspective and basis for empirical analysis, not a set of ready-made hypotheses. Thus, rejection of a lower-level hypothesis need not imply rejection of the systems approach.

This apparent irrefutability of the systems approach in no way invalidates its use. Most theoretical approaches incorporate variables and relationships at higher levels that are not directly observable, but which have operational correlates at lower levels; the relevance or otherwise of the systems approach will be debated on the basis of the performance of lower-level hypotheses.

As far as the empirical hypotheses consistent with alternative theoretical approaches are concerned, hypotheses may be similar or conflicting between these different frameworks, or it may be that a hypothesis may have no corresponding or conflicting hypotheses in other approaches. The latter is particularly likely if one approach suggests a rich vein of testable lower-level hypotheses.

These possibilities are of particular relevance as far as the regression analysis conducted in Chapter 7 is concerned. Some of the hypotheses may be consistent with the neoclassical theory of the firm, while with others it is difficult to see how they might be developed in a project-based theory of the firm framework. It will be argued that the systems approach must be judged on assessment of the lower-level hypotheses considered *as a whole*. The regression analysis of Chapter 7 appears to offer a good explanation, not only of the behaviour of R & D at industry levels, but also characteristics of uptake of basic research activity by corporations. Although conventionally it is generally assumed that resource allocation to this latter phenomenon is particularly difficult, if not impossible, to account for in economic analysis, Chapter 7 suggests that this may not be the case. Not all the lower-level hypotheses are supported by the evidence in this particular study, but it will be argued that the general performance of the regression analysis is good.

The analysis is therefore intended to provide a useful though *partial* framework for the analysis of aspects of technological change in the corporation. One area of obvious relevance which is virtually neglected is that of selection of projects and resource allocation within the R & D budget constraint; however, it is argued later that this is justified in terms of the points made in Chapters 4 and 5. Bearing in mind such potential restrictions on the analysis, a main aim of the book is the provision of a useful conceptual framework for the analysis of resource allocation to technological change in the large modern corporation, and demonstration of the potential applicability of *rational* analysis to areas where it has been frequently suggested to be irrelevant.

However, there is a second main objective of this book which evolved from, and was stimulated by, the nature of the problems encountered in developing this framework. It will be argued that the emphasis on individual *elements* in the resource-allocating process in analytic theories such as those of the neoclassical economists may actually obscure or inhibit understanding of decision-making in the corporation. Neoclassical theory has as its building-blocks the individual consumer, product, and projects, larger units being defined in terms of sums of individual elements. The technique of aggregation is used in moving from micro-levels to higher levels of analysis.

It is a basic tenet here that reduction to component elements and definition of higher levels as aggregates may be inappropriate in certain circumstances and for investigation of specific phenomena. Instead, a systemic or *holistic* view of the corporation is suggested as an alternative approach to aspects of resource allocation. Two main reasons are given for this. Firstly, advantage may be taken of redundancy at lower levels; only limited and highly abstract detail may be necessary to describe or approximate adequately a system's behaviour. This is illustrated in a pragmatic way by simulation models which in certain circumstances may provide a good description of the behaviour of complex systems while utilising only highly selective and schematic information. Secondly, system description and behaviour may be non-reducible as far as specification in terms of constituent elements is concerned. Pattern, configuration or 'Gestalt' may be established at relatively high levels of abstraction, and may not be directly derivable from consideration of components alone. Both these arguments are used in interpreting the corporation as an adaptive, hierarchically structured system, and will be developed at greater length in subsequent discussion.

Thus the objectives of the present analysis are basically twofold. Firstly, it is intended that a satisfactory and useful approach to some decision-making problems relating to technological change in the modern corporation may be developed. Secondly, it is hoped that a convincing case may be made for the argument that the dominance of a reductionist perspective in standard economic approaches may hinder rather than assist model-building in some cases. It is hoped that the following analysis may demonstrate the possible usefulness of an alternative economic theory of the firm based on the constructs of general system theory.

2 Industrial Research and Development at Project Level in the Modern Corporation

The existence of technological change has created numerous theoretical and empirical problems for economists. It directly challenges a basic assumption of neoclassical microeconomics that the state of technology is given; invention alters the production function of firms and/or the types of products produced, and the process by which such development occurs is as yet little understood. Firstly, the generation of radical innovation may involve complex and highly technical issues which are difficult to communicate from scientists and technologists to economists. Secondly, even if the problems of communication between disciplines could be solved, there is little evidence that professional R & D workers could articulate meaningful and useful models of R & D activity, as we hope to demonstrate later in this chapter.

The inherent difficulties for model-building in this area are reinforced by the role of R & D as a peripheral industrial activity until relatively recently. Economic analysis has tended to concentrate on areas more amenable to treatment using its sophisticated and quantitatively based techniques, and when research and development (R & D) could no longer be ignored owing to its rapid post-war increase in developed countries, a number of approaches tested and accepted in other areas proved difficult, if not impossible, to apply to R & D work. This will be discussed further in Chapter 3, and specific approaches will be examined in this respect. However, before we can do so, it is necessary to analyse why and how research and development may present particular problems for corporate decision-makers, and this is the purpose of this chapter.

With this aim in mind, the prime concern of the present chapter is with the effect of uncertainty on the applicability of rational models and analysis. First of all, we shall examine the types of activity associated with R & D work and associated classificatory problems. The difficulties of applying rational analysis in these areas will be discussed, and evidence for the central importance of uncertainty presented. The effect of uncertainty on output of types of R & D activity as well as in the relationships between types of R & D activity will then be examined in attempting to illuminate its role in the R & D process.

It will be suggested that *at project level* there is as yet little evidence that rational models of technological innovation exist or are obtainable, which provides an essential basis for the argument developed in Chapter 3. However, in Chapter 4 we shall point out that it is possible to identify at least *two* separate areas of R & D decision:

(i) the determination of the overall budget,
(ii) the allocation of resources to individual projects;

and suggest in later chapters that the inapplicability of rational models to the latter area may not necessarily mean that rational analysis is of little use in the determination of the overall budget for technological innovation.

Firstly, however, we shall discuss the nature of invention and innovation, and what is implied by those R & D activities, as well as discussing the possible relationship between science and technology in the R & D process.

INDUSTRIAL INNOVATIVE ACTIVITY: CONCEPTS AND CLASSIFICATION

Innovative activity by the firm may be interpreted as being, in general, of two distinct types: innovation-generating and imitative behaviour. As Nelson (1972) points out, for a model (of firm behaviour) really capable of generating and responding to technological change it seems essential to incorporate 'some kind of an innovating or internal search mechanism for improvement, and some kind of an imitation mechanism whereby what one firm does can induce another firm to do likewise' (p. 44).

In the following chapters we shall be primarily concerned with the internal search for innovations, or innovation generation, rather than with the imitation or diffusion process. However, in the development of a framework we shall also consider how the allocation of resources to the external search for innovations may be included in a more general analysis.

Internal search activity for technological innovation is the responsibility of corporate research and development (R & D) departments. While definitions and interpretations of R & D activity vary, those of the National Science Foundation (N.S.F., 1973b) are widely accepted; research and development is defined by the Foundation as: 'Basic and applied research in the sciences and engineering and the design and development of prototypes and processes'[1] (p. 19).

The Foundation also provides standard definitions of the types of R & D activity mentioned above. Basic research is defined as: 'Original investigations for the advancement of scientific knowledge not having specific commercial objectives, although such investigations may be in fields of present or potential interest to the reporting company' (p. 19).

Basic research is distinguished from applied research,[2] the latter being defined as: 'Investigations directed to the discovery of new scientific knowledge having specific commercial objectives with respect to products or processes. This definition differs from that of basic research chiefly in terms of the objectives of the reporting company' (p. 19).

Development work is defined as: 'Technical activities of a non-routine nature concerned with translating resource findings or other scientific knowledge into products or processes. Does not include technical services or other activities excluded from the above definition of research and development' (p. 19).

The N.S.F. definitions are primarily concerned with the presence or absence of commercial objectives. This is a fairly restricted view in the light of recent development in the theory of the firm.[3]

A further problem area is the distinction between basic and applied research. In the N.S.F. definitions it tends to parallel that usually made between scientific discovery and technological invention respectively. However, if the definitions are strictly applied, scientific discovery oriented towards commercial application would be interpreted as applied research, while, as Schmookler (1962a, p. 45) points out, some inventive work would be interpreted as development. 'Discovery' and 'invention' are often difficult distinctions to make in practice, but Siegel (1962) provides a distinction consistent with the consensus in the literature: 'discovery may be a "new" fact, principle, hypothesis, theory or law concerning natural (including human) phenomena that are observable directly or through their effects' (p. 442). Invention, according to Siegel, 'may be regarded as purposeful and practical contriving based on existing knowledge (theoretical or applied) and uncommon insight or skill; that is as the act of bringing to workable condition a potentially economic or usable process or product . . . that has a significantly novel feature' (p. 442).

While the distinction between scientific and technological R & D may be difficult to make in practice, Price (1965), for example, suggests that the two bodies of knowledge are quite separate, both conceptually and in their historical development, as would be expected from Siegel's distinction. One obvious distinction from an economic point of view is that scientific knowledge is non-patentable. Free and public communication of results is a traditional and ingrained value of the scientific community, and consequently it is likely that the firm will not be able to gain the full economic value created by research for new scientific knowledge (Nelson, 1959a). However, recognition is generally made that property rights in invention are necessary at least to some extent in order for technological research to be conducted by private companies. Non-appropriability of results would result in the potential external economies of research being high, but with the private incentive to conduct research being reduced because of the ability of other firms to copy or imitate expensively researched innovations. Consequently, in the light of the separate traditions of science and technology, science is frequently regarded as both the natural and rightful concern of social policy rather than private industry (Nelson, 1959a, pp. 299, 306). This is supported by the observed distribution of R & D activity among sections in the United State: N.S.F. statistics indicate that private industry carried out 14.2 per cent of basic research, 57.2 per cent of applied research and 85.5 per cent of development conducted by all sectors of the U.S. economy in 1972 (National Science Foundation, 1972a, table pp. 6–7). Universities and colleges were the main performers of basic research (56.7 per cent) with the residual being accounted for by federal and federally funded institutions and 'other non-profit institutions' (29.1 per cent).

However, distinction between the two systems can also be made in terms of the habits and mores of scientists and technologists. Price (1975, p. 130) suggests that scientists are motivated towards discovering new scientific knowledge for reasons of prestige and recognition of professional excellence, while technologists are concerned with more commercially oriented work concerned with the search for privately appropriable invention.[4] According to Price, scientists and technologists are generally very different types of people with differing motivation and even training, and Gibbons and Johnston (1974, p. 223) suggest that science and technology may be considered distinct to the extent that it is possible to identify two separate professional groups distinguishable in terms of their respective attitudes, values and norms. Elsewhere, Price (1969, pp. 171–2) states that scientific effort is directed towards publishing results, while technological effort is directed by the opposite motivation of concealment, at least until patenting is achieved.[5]

A pervasive difficulty, whatever the basis for distinguishing between R & D subsystems, is the difficulty of separating them out in practice. A UNESCO report (1970, in Freeman, 1974, pp. 358–9) emphasises the conceptual usefulness but empirical difficulty of distinguishing between R & D subsystems. An O.E.C.D., D.A.S./S.P.R. report (1970) puts the problem in context by pointing out that the appropriate allocation of an R & D activity to one of the categories may be neither natural nor obvious. Nelson (1959a, p. 300) emphasises the 'fuzzy boundaries' between conceptually distinct categories of R & D, and Schmookler (1966) suggests that the process of technological invention may involve the simultaneous synthesis of various aspects of science and technology, the implicit sequentiality of the basic research/applied research and development categories tending to mislead interpretations of the innovative process. Also, Falk (1973, p. 188) and Salomon (1971, p. 11) emphasise the difficulties of distinguishing the traditional categories of R & D at both conceptual and operational levels, while Gibbons and Johnston (1974, pp. 223, 241) emphasise the close similarity of scientific and technological methods of investigation, and the wide variety of potential forms of interaction between the two forms of research.

There are therefore a number of different types of R & D activity which may combine and interact in the course of particular projects. As we shall see in the next section, a number of past models have assumed the form of interaction to be linear or sequential, i.e. a development proceeds in an orderly fashion from one stage to the next. We shall now examine this assumption, together with the more general proposition that rational modelling is feasible in highly innovative activity.

THE RATIONAL MODEL OF INNOVATION

Schon (1967) and Gold (1971) have identified, and strongly criticised, the pervasive and widely held view that innovation is a rational process which is capable of management, direction and control.[6] While rationality is not a simple concept to define, Simon (1965) suggests a multidimensional interpretation, providing in particular an important distinction between objective and subjective rationality:

> . . . rationality is concerned with the selection of preferred behaviour alternatives in terms of some system of values whereby the consequences of behaviour be evaluated. . . . a decision may be called 'objectively' rational if *in fact* it is the correct behaviour for maximising given values in a given situation. It is 'subjectively' rational if it maximises attainment relative to the actual knowledge of the subject. (pp. 75–6)

Whether rationality is subjective or objective, it implies that the decision-making process is goal-directed, manageable and controllable. Schon (1965, pp. 3–5, 19–20) cites evidence which suggests that innovation is widely regarded as being a rational process, or, if not, capable of being such. Gold (1971) also provides extensive evidence to suggest that there is a 'synoptic model' of innovation (see pp. 213–15), in which rationality of decision-making constitutes the definitive theme. The synoptic model has four major building-blocks:

(a) the belief that technological innovation is inherently attractive in corporations, particularly in the context of potential economic rewards;
(b) the belief that technological innovations are planned and controlled by management;
(c) the belief that decision-making is rational with built-in evaluative feedback loops;
(d) the belief that R & D constitutes the most important means of effecting growth and profitability.

Influential proponents of the rational view include Machlup (1962b) who suggests (p. 153) that the process of invention has become systematic, routinised and amenable to rational analysis, and Schumpeter (1954) who states:

> . . . it is much easier now than it has been in the past to do things that lie outside familiar routine – innovation itself is being reduced to routine. Technological progress is increasingly becoming the business of teams of trained specialists who turn out what is required and make it work in predictable ways. The romance of earlier commercial adventure is rapidly wearing away because so many things can be strictly calculated that had of old to be visualised in a flash of genius. (p. 132)

However, Schon provides evidence, and Gold cites an impressive array of studies, to infer that the rational view of innovation and the constituent 'building-blocks' of the synoptic model are misleading. Extensive empirical evidence is supplied by Jewkes et al. (1969) and Langrish et al. (1972). Jewkes et al., concentrating on the invention stages of the innovation process, found from 60 case studies that invention is extremely difficult to direct or control consciously (pp. 108–9), that intuition and chance were central factors in technical progress (p. 169) and that prediction is practically impossible as far as most inventions are concerned (pp. 170–7). Langrish et al., concerned with the total innovation process in 36 studies of Queen's Award winners, emphasise the plurality of sources of innovation and the unreasonableness of using or adhering to the concept of a linear process of innovation development (p. 7).[7]

Price and Bass (1969) identify a conceptual model of the innovation process based on assumptions of rationality and linearity similar to those criticised by Gold and Schon, and argue that for radical innovations the organisations which introduce them must undergo major internal development and change, much of which cannot be programmed in advance. This has subsequently been supported by empirical research by a team from the Illinois Institute of Technology (1968), and by Globe et al. (1973), both of which found that non-mission-oriented research[8] played an extremely important role in the development of selected major innovations. The former study (codenamed TRACES) found that 70 per cent of events considered important in leading to the obtaining of the five major innovations studied were non-mission-oriented, while Globe et al. found that non-mission-oriented research accounted for 57 per cent of 'significant events'[9] in the pre-innovative period (the period before the original idea of the innovation is conceived), 16 per cent of 'significant events' during the innovative period (from conception to realisation of the innovation), and 10 per cent during the post-innovation period (marketing, diffusion and improvements period).[10]

As would be expected in such circumstances, estimates of development time or cost for any particular project are typically subject to substantial error. Marshall and Meckling (1962) found in a study of 22 major military development projects that the mean value of the ratio of the most recent available final cost estimates to the earliest available estimates was between 2.4 and 3,[11] and extensions of development time by one-third to one-half to be typical (p. 474). Significantly for the predictability of innovation cost, various around the mean value was also high (see p. 469), tending to rebut interpretation of the mean value of latest to earliest cost estimates as being due simply to optimistic bias.[12] As would be expected from criticisms of the rational model, high variability and unreliability is typically the case in major and radical innovation developments.[13]

In this context, normative analysis of research and development management has frequently emphasised the inability of the decision-maker to impose a high degree of control over the innovation process. Nelson (1962) noted the tendency for selection of projects in a highly productive industrial research laboratory[14] to be effected by an evolutionary or 'natural selection' process (p. 572) and has advocated that R & D should be explicitly recognised as being unreliable and unpredictable and that, especially in the early stages, a high degree of control should not be imposed over the innovation process; instead a number of projects should be permitted and encouraged to develop, and a high rate of project failure should be anticipated and accepted in advance (Nelson, 1959b, 1972). While Gold (1971) does not offer simple normative solutions to his criticisms of the rational model, he does suggest that analysis should concentrate on how the process of innovation actually operates rather than trying to mould the planning and decision-making process according to preconceived analytical frameworks (see particularly p. 245). Schon (1967) argues that technological change is irregular and unpredictable, and that rather than attempt to mechanise or routinise change, an atmosphere receptive and conducive to change should be created in the corporation.[15]

Klein (1962),[16] basing his argument on the Marshall–Meckling results among other case studies of innovation development, infers that efficiency in the 'narrower sense' of the rational model inhibits improved decision-making in R & D (p. 497), and that treating R & D projects as if they are liable to be controllable and well behaved may turn out to be costly in the end (p. 496). Like Gold, Klein suggests that the innovation process may not be analysable in terms of the standard optimisation techniques which are applicable to routine projects. Klein argues that multiple approaches and imposing low constraints on the progress of innovation development will provide greater opportunity for the development of a viable and useful system.

Hirschman and Lindblom (1962) suggest that, rather than being an isolated viewpoint, Klein's work along with joint work with Meckling may be identified as one aspect of convergent thinking on normative decision-making in the area of economic development and public policy-making as well as technological R & D itself.[17] Hirschman and Lindblom identify a number of specific points of agreement between the various interpretations, particularly the likelihood that maximising techniques for analysis may prove positively detrimental in many areas of decision-making and policy formulation, that 'disorderly' development of decisions, innovations and policies may be desirable, and that elaborate and extensive attempts to specify alternatives, identify end-goals and integrate planning are frequently counterproductive in conditions of high complexity, insufficient inducements to decision-making, uncertainty and/or limited decision-making capacity.

As Hirschman and Lindblom suggest, the respective sources place different emphasis on the importance of each justification for abandoning the rational model. They point out that the rationale for criticising the applicability of the rational model in Klein and Mecklings's view is *future uncertainties*, i.e. the inability of decision-makers to anticipate or predict the shape of future technological developments .[18] It will be suggested in the next section that uncertainty is the dominant consideration in analysis of the innovative process, and that it is the influence and implications of uncertainty for decision-making that have to be focused on in constructing models of innovation.

UNCERTAINTIES IN THE INNOVATION PROCESS

Knight (1921) first distinguished between measurable and quantifiable uncertainty or risk, and unmeasurable or true uncertainty. Conceptually, risk and uncertainty are quite distinct, although in practice they may be difficult to separate. Risk depends on the existence of replicability and homogeneity of events, and the consequent calculation of probabilities of occurrence, cost, etc., using statistical techniques. True uncertainty on the other hand cannot be reduced to objective probabilities and consequently qualitative analysis involving 'hunches', intuition and judgement tends to play a prominent role in decision-making when uncertainty is present, and when decisions are totally or partially non-programmable. There is in fact a long-standing and continuing dispute between Bayesian and non-Bayesian statisticians as to the legitimacy of quantification in unique decisions. We shall touch on this problem in the next chapter, but for the moment we shall utilise the non-Bayesian distinction between risk and uncertainty.

Essentially, the existence of risk poses no real problem for the rational model. Given the existence of objective probabilities of future states, certainty equivalents may be calculated for specific courses of action, and rational choice of alternatives may be made, given the objectives of the decision-maker and perceived values and estimated probabilities of possible outcomes. However, in conditions of uncertainty the assumptions of rational analysis no longer hold (specifically those requiring that all alternative future states be specified, with associated values and risk of occurrence); consequently the rational model itself may no longer be applicable, as Gold and Schon have shown, if uncertainty is a significant factor affecting the decision-making problem.

In fact, uncertainty is widely regarded as an integral and effectively definitive aspect of innovation, and decision-making for innovation. As Gold (1971) points out:

> the overwhelming evidence from empirical studies so far is that, except for relatively routine improvement projects, unpredictability is pervasive. It seems to be difficult to predict: the kinds of inventions or discoveries likely to occur; the kinds of applications likely to be made of new discoveries; how close to success given undertakings are; and even how alternative designs or carefully developed theoretical models will turn out. (pp. 217–18)[19]

Freeman (1974, pp. 223–7) identifies three broad categories of uncertainties which are associated with innovative activity: general business, market and technical uncertainties. General business uncertainty applies to all decisions which relate to the future, to the extent that they may be influenced by environmental variables (political, legal, economic, etc.). Since innovations typically have a longer gestation period than other investment decisions, it is particularly relevant to this decision class. The other two forms of uncertainty are project-specific and according to Freeman are not insurable against as risk. Technical uncertainty refers to realised standards of performance under various operating conditions for a given expenditure on R & D, while market uncertainty refers to the extent to which the innovation will be commercially successful for a given product specification. Each category of uncertainty is likely to be of importance for all but minor development projects: as Hamberg (1963) points out, unforeseen problems commonly arise that were not anticipated in preliminary investigations.

This is of course as we would expect from the TRACES and Globe et al. studies, given the importance of non-mission-oriented research in the development of the selected innovations. However, the technical uncertainty surrounding innovation projects varies qualitatively according to the type of project. As Bowie (1963) points out, 'As one progresses from research through development, design and pilot production into full production, he is passing from a state of low predictability of specific accomplishment to one of high predictability' (pp. 280–1). Bowie states that basic research 'by its very nature' is unpredictable of specific accomplishment (p. 281). Freeman (1974. p. 226) similarly identifies qualitative differences in uncertainty for various categories of innovations (see Table 2.1), identifying degree of uncertainty as being associated with the original source of the innovation. Fundamental

TABLE 2.1 Degree of uncertainty associated with various types of innovation

1. True uncertainty	Fundamental research Fundamental invention
2. Very high degree of uncertainty	Radical product innovations Radical process innovations outside firm
3. High degree of uncertainty	Major product innovations Radical process innovations in own establishment or system
4. Moderate uncertainty	New 'generations' of established products
5. Little uncertainty	Licensed innovation Imitation of product innovations Modification of products and processes Early adoption of established process
6. Very little uncertainty	New 'model' Product differentiation Agency for established product innovation Late adoption of established process innovation in own establishment Minor technical improvements

Source: C. Freeman, *The Economics of Industrial Innovation*, (Penguin Books, London, 1974) p. 226.

research, and by implication basic research, is an extremely uncertain undertaking, while applied research, if the N.S.F. definitions are to be interpreted literally, is interpretable in terms of categories 1–3 (and possibly 4) with associated high degrees of uncertainty. Development work (categories 4–6) on the other hand has relatively low levels of uncertainty associated with it.

The degree of uncertainty surrounding the eventual value of R & D work may therefore depend on the type of R & D work being carried out. This is a generally recognised and accepted phenomenon, yet one problem is that implicit in such recognition appears to be the idea of a linear model (see, for example, Bowie's idea of 'progression' above). We have already cited evidence to suggest that simple, continuous progression from basic research through intermediate stages to production is not typical of the R & D process, and indeed this is recognised in some models of the R & D process developed in recent years.[20] It would help to clarify understanding as to the role of uncertainty in the R & D process if this apparent inconsistency could be resolved; if the linear model is discredited, then on what basis is it possible to discriminate in terms of degree of uncertainty associated with a particular type of R & D activity? In the linear model, the further the appropriate stage or activity is removed from actual innovation, the longer the time and the greater the number of intermediate stages to development completion, and consequently the greater the degree of technical, market and general business uncertainty associated with the activity. The abandonment of the linear model means that we have to establish an alternative basis for assessment of the relationship between different kinds of R & D activity, since one area of concern in Chapter 7 is the distribution of resources within the R & D budget. Therefore in the next section we look at the relationship between different kinds of R & D subsystems; however, before we do so, the characteristics of respective R & D subsystems must be examined more closely, in particular the role of uncertainty *in each subsystem.*

We shall generally interpret basic research in the N.S.F. definition as being synonymous with scientific research, and applied research as being synonymous with technological research (or product- or process-directed applied science). This is consistent with the use of the terms by Price and Bass (1969), but is not an entirely happy interpretation in the context of the N.S.F. usage. This is because the N.S.F. definitions emphasise objectives, while ideally we would prefer an interpretation that deals with the actual informational output of each subsystem (such as in Machlup's framework, 1962a, pp. 180–1). We retain the N.S.F. definitions because N.S.F. data will be analysed in Part II.

While bearing these qualifications in mind, conceptually at least subsystems may be simply distinguished from one another. However, references to the level or order of uncertainty associated with R & D work tend either to emphasise the high level of uncertainty typically associated with R & D work in general,[21] or to analyse the level of uncertainty associated with the inputs of a particular stage as far as their expected contribution to final output is concerned.[22] There is little evidence of attempts to analyse systematically the level of uncertainty between inputs of a particular subsystem and the anticipated intermediate output of the subsystem. Such analysis, together with specification of the relationship between a particular subsystem and final output, is essential if the R & D process is to be adequately examined.

In fact, thc studies that have been carried out at subsystem level tend to report unpredictability and uncertainty within subsystem boundaries, particularly in the case of 'basic research' and 'applied research' subsystems. Price (1964, pp. 199, 201), Taton (1957) and Kuznets (1962, pp. 20–1) emphasise the pervasive uncertainty associated with basic research,[23] and researchers identifying similar properties of applied research include Griliches (1962, p. 352), Kuznets (1962, pp. 21–2), Gold (1971, p. 224), Schon (1967, chaps. 1–2) and Price and Bass (1969, p. 805).

Some studies suggest that these subsystems are comparable in this respect. Jewkes et al. (1969, pp. 101–2) and Taton (1957) point out the role of chance in the realms of discovery and invention, while Siegel (1962, p. 443) notes the unpredictability of both scientific discovery and technological invention. Price (1965) suggests that attempts by society and industry to improve direction and control on both the search for scientific knowledge and technological developments are almost certainly doomed to failure in the former case, and possibly also in the latter (p. 564).

Development work is generally regarded as less uncertain than the other two categories of R & D activity, though the activity itself may vary widely in the demands it makes on researchers.[24] Carter and Williams (1959, p. 48), Schmookler (1962a, p. 45), Jewkes et al. (1969, p. 28) and Freeman (1974, p. 229) testify to the relatively low degree of uncertainty associated with development compared to the other subsystems. However, uncertainty is still a feature of development, even if the degree of uncertainty is low compared to the basic research and inventive work subsystems.

The uncertainty of 'basic research' and 'applied research' might also be expected on *a priori* grounds. Basic research is concerned with the search for new scientific knowledge, and consequently the eventual output must by definition be at least partially unknown to the decision-maker when resources and inputs are allocated to a particular project, otherwise the allocation of resources to such activity would be redundant and wasteful.

If the output is uncertain, so also will be the anticipated value from allocation of resources to a project. In a similar manner, inventive work is concerned with the creation of novel products and processes, the level of uncertainty of each being related to the extent to which the invention sought is radical and different from those preceding. As with basic research, the eventual output is to a greater or lesser extent unknown and consequently uncertain. If such uncertainty could be interpreted as measurable uncertainty or risk, there would still be opportunity for the application of quantitative optimising models for R & D allocations; unfortunately, as far as the use of such techniques is concerned, use of risk presumes that all events in the class of events to which it refers are homogeneous as far as the elements of the events deemed relevant to the analysis are concerned. Yet if discoveries and inventions are analysed in this manner by identifying a 'unit of discovery' or 'standard invention', the essential quality of discovery and invention is effectively negated, since they are defined in terms of differences from preceding science and technology respectively, not similarities (see Freeman, 1974, p. 225).[25] The extent to which a discovery or invention may be regarded as being radically different from those preceding depends on circumstances, yet since they are defined as non-homogeneous and unique events (at least as far as individual or institutional decision-making is concerned), the implication for the basic research and inventive work subsystems is that true uncertainty must be regarded as an integral feature of those activities.

Development work also may be regarded as being concerned to some extent at least with novel problems in each project. Ames (1961, p. 373) and Machlup's framework (1962a, pp. 180–1) both suggest that practical problems or 'bugs' may arise at this stage, indicating that uncertainty may still be present. However, development work, while it may be highly uncertain if new design and testing techniques are required, frequently follows well-tried and familiar ground rules for the design and development of innovations, and consequently it may involve lower degrees of uncertainty than the other subsystems. Similarly with 'new-type plant construction', an element of novelty and uncertainty is still present in the system.

Consequently each subsystem may be regarded as innately uncertain, the degree of uncertainty of each subsystem being related to the novelty of the task required. However, there are two other aspects of R & D activity which are of importance in considering the reliability and predictability of eventual economic value of inputs in a particular subsystem: (i) the means by which a particular subsystem is coupled to final output by other subsystems; and (ii) uncertainty pertaining to linkages and relationships *between* subsystems. We shall attempt to relate these concepts to the N.S.F. framework.

HIERARCHICAL ARRANGEMENT OF SUBSYSTEMS

The first problem area mentioned above, coupling of subsystems, is of relevance in establishing the inputs into a subsystem and the anticipated economic reward to these inputs. As far as the relationship between science and technology is concerned, Price (1975) relates this to the social, cultural and economic separation of the two subsystems, referred to earlier; 'In normal growth science begets more science and technology begets more technology' (pp. 129–30).

This suggests that technology has minimal input from science under normal conditions, and that the two subsystems are, at best, only loosely connected. As far as links between these research subsystems and development are concerned, while applied research may carry forward into development the scientific knowledge inputs required for development word, basic research has more tenuous links with development, typically requiring applied research before it serves as useful input:

> Science enters innovation already embodied in technological form. It may be relatively rare for a piece of curiosity-oriented research to generate a piece of new technology, but once the process has occurred the technology can be used over and over again and developed into more advanced technology. There seems to be much justification for the view of Price that technology builds largely on earlier technology. (Langrish et al., 1972, p. 40)

The embodiment of scientific research in applied research, and the role of development work in converting inventions into detailed construction and design plan, suggest the idea of progression from higher stages to lower stages, in that order.[26] Machlup's framework referred to earlier is useful in this respect, in that not only does it indicate such a progression (basic research → inventive work → development), but it also recognises that the progression may be erratic and discontinuous, jumping from a particular stage to a higher or lower one, or to a different activity in the same stage.

Thus, although progression through lower stages will be necessary for a project operating at a particular stage, it may not be sufficient. Projects may be referred back to earlier stages as 'bugs' appear. In fact it is useful to regard each subsystem not as links in a linear chain, but as forming levels in a hierarchy of R & D subsystems. The higher its level in the overall system, the more remote the informational content is from final output. Thus, the intermediate output of basic research generally requires embodiment in a new technological concept, and design and development work before the stage of tooling-up for production can be reached. However, while lower stages in general are necessary intermediary stages between higher stages and final output, redirection of an R & D project or idea to a higher stage is always possible depending on unexpected difficulties, new opportunities or revised priorities.

While the idea of hierarchy of these subsystems must of necessity be a rather crude approximation to reality, given the rather messy and complex nature and relationships of these subsystems, emphasised earlier, it may be interpreted as an improvement on the traditional concept of a unidirectional linear model. This concept of a hierarchical ordering of the R & D subsystems will form a central part of the argument in subsequent chapters.

LINKAGES BETWEEN R & D SUBSYSTEMS, AND PROJECT UNCERTAINTY

The second problem area relevant to consideration of uncertainty surrounding economic potential of inputs into particular subsystems is that relating to the linkages between respective subsystems. Earlier, consideration was given to relationships between inputs of a subsystem and outputs of the subsystem; in a sense the problem here is the reverse, since we are concerned with how the output of one system relates to the input of the lower subsystem. Even if each subsystem could be regarded as determinate with respect to production of its own particular intermediate output, once input of a higher system is related to anticipated effect on output of a lower system, opportunity for misdirection of effort or uncertainty of input value may arise owing to mismatching or poor integration of subsystem linkages.

Of particular interest are the linkage characteristics between inventive activity and the adjacent subsystems. As discussed in the previous section, as far as Price (1975) is concerned, science and technology are two loosely connected 'twin' systems: 'Only rarely, but then dramatically and making a historical mountain peak, do the twins show a strong interaction' (p. 130).[27]

In general, links between science and technology are regarded as indirect, unreliable and usually highly lagged, with technology building on 'old' science rather than actively interacting with current scientific activity. Schmookler (1962b, pp. 224–5) found little evidence of active links between scientific and technological subsystems in a study of selected important inventions, while Langrish et al. (1972, pp. 39—42), Jewkes et al. (1969, pp. 21, 199–203) and Ben-David (1972, pp. 184–5) similarly identify the absence of strong, direct links between science and technology. To a large extent, this is regarded as a consequence of the separate scientific and technological 'cultures' mentioned earlier. As far as the impact of basic research on development is concerned, generally extensive scientific and technological research remains to be done on 'pure' research findings before the stage is reached of actually directing work to specific product/process development. Orientation of basic research output into applied scientific and technological 'channels' is usually necessary.

A frequently cited example of 'pure' basic research leading eventually to commercially viable innovation through applied research is that of the development of nylon.[28] In 1927 E. I. du Pont de Nemours & Co. decided to fund a programme of fundamental research, led by W. H. Carothers, on a yearly budget of $250,000. The goal of this programme was to 'discover scientific knowledge regardless of immediate commercial value' (Mansfield, 1968a, p. 49). The results of the programme were only of academic value for the first three years, until a superpolymer was discovered with a high degree of flexibility and strength, though since it was weakened by hot water, it was of no direct use commercially.

However, this discovery served as a stimulus in directing the programme towards synthesising related compounds in the hope of discovering a commercially viable fibre. This point might therefore be selected as the beginning of 'applied research' according to the N.S.F. interpretation. Yet it took some time before a commercially promising superpolymer, '66' polymer, was discovered. Carothers at one point had in fact abandoned his search for a commercially useful compound and only returned to the project after encouragement by a new director of du Pont's Chemical Department. It took eight years from the discovery of the superpolymer which prompted the applied research phase of the project to the point where '66' polymer (nylon) was first commercially produced. Substantial applied research had been required before the initial discovery of a superpolymer with possible commercial properties.

MacLaurin (1953, p. 99) points out that pure science rarely leads *directly* to a patentable invention, while Nelson (1959a), in an analysis of the economics of basic research, suggests that this progression of basic (fundamental) research through applied research and then development tends to be the rule in industrial R & D: 'significant advances in scientific knowledge are often not directly and immediately applicable to the solutions of practical problems and hence do not quickly result in patents. Often the new knowledge is of greatest value as a key input of other research projects which, in turn, may yield results of practicable and patentable value' (p. 302).[29]

Project Hindsight, which was a study designed to identify the 'events' leading to the development of new post-war weapons systems in the United States, found little relation between undirected science (basic research) and the development of technological innovations. Only 0.3 per cent of 'events'[30] were identified as undirected science. It was concluded on the evidence of the study that 'the process by which science moves into technology and utilisation . . . is clearly not the simple direct sequence taught by the folklore of science' (Scherwin and Isenson, 1967, p. 1576). Further, Sherwin and Isenson found that generally many significant technological innovations were required to develop a viable new system, suggesting the importance of applied research in this sphere. Links between basic research and development, where they existed, were likely therefore to be buffered by an intermediate stage of applied research.[31]

However, links between applied research and development input are also significantly unreliable. Nelson et al. (1967, p. 95) state that studies of patent utilisation in the United States suggest that approximately 50 per cent of patents are eventually used. They state that since the bulk of R & D expenditures of a project is concentrated on the design and development stage, many are abandoned before commitment to development is undertaken. This would suggest a bias in the development projects undertaken towards the less uncertain, and consequently concentration of analysis on only patents and non-patented inventions carried to the point of development would tend to obscure a significant level of technical and market uncertainty remaining after the patenting stage has been reached.

Although studies of R & D activity tend to report heavy concentration on development expenditure relative to research,[32] these understate the commitment of resources required at the development stage. Many projects fail to be approved for full-scale development because of remaining uncertainty, and the anticipated resources and time required, to get the project on the market. Consequently, retrospective evaluations of development costs as a proportion of total R & D costs of individual completed projects would indicate even more strongly the relative expense of development compared to research. Expensive and time-consuming development operations such as design, construction and evaluation of prototypes and pilot plants add to development cost of innovations,[33] magnifying the expenditure incurred in developing the typically relatively inexpensive research projects. Not surprisingly, the route from applied research to development is seldom an automatic one, revision and elimination of projects being common between stages.

Therefore, there typically exists considerable imperfection and uncertainty in the output–input linkages between subsystems, as well as in the input–output characteristics of subsystems themselves. The implications for a project operating at a particular stage or level in the R & D system hierarchy are that not only does the reliability or otherwise of subsystem input relative to final output depend on subsystem operations themselves, but it is also contingent on the degree of determinacy of links between subsystems. Uncertainty relating to perceived economic value of inputs at a particular stage will be a consequence not only of uncertainty in input–output relations of that and lower stages, but will also be magnified by uncertainty pertaining to informational output exchanges between subsystems.

At this point some general points can now be summarised regarding R & D subsystem operations:

(a) R & D subsystems may be regarded as forming a hierarchy in the R & D system, stages in the hierarchy being determined by distance from final output. In the context of the N.S.F. definitions, the hierarchy proceeds from basic research to applied research to development to final output.
(b) Inputs and operations of a particular subsystem have associated uncertainty with respect to intermediate output of that system, while uncertainty is also a general characteristic of subsystem linkages. Uncertainty pertaining to subsystem input depends on degree of novelty of anticipated intermediate output.
(c) Input into a subsystem at a particular level in the hierarchy generally has to be processed through the subsystem and lower subsystems before it contributes to final output.
(d) Consequently the higher the level of R & D subsystem operation, the greater the degree of uncertainty typically surrounding subsystem input, with respect to anticipated contribution to final output. Anticipated economic value of subsystem input is affected

> not only by inherent uncertainty of the subsystem itself, but by uncertainty surrounding operations of lower subsystems and subsystem linkages.

Freeman's analysis of degree of uncertainty typically associated with different types of innovation (Table 2.1) is interpretable in such terms. For example, fundamental research has complex and indirect links with innovation since inputs into that system generally require processing by each of the four subsystems in Table 2.1, whether in original or in some mutated form. Also, linkages between fundamental research inputs and other subsystems have already been pointed out to be loose and tenuous, and consequently fundamental research is a highly uncertain undertaking for the firm.

On the other hand, obtaining an innovation from a patented idea involves comparatively little uncertainty, inputs into the development stage being fairly close to final output if no problems or hitches arise. Consequently, while allocation of R & D activity to one or other subsystem may be difficult or arbitrary,[34] the concept of hierarchical arrangement and relationship of subsystems facilitates analysis of the characteristics of inputs and operations for particular subsystems, and permits investigation of the degree of uncertainty associated with a particular project by interpreting the R & D system as being composed of a series of interrelated but poorly integrated subsystems.

SUMMARY

Research and development has been interpreted here as a complex of interrelated subsystems arranged hierarchically, each inherently uncertain, but with uncertainty being compounded as projects concentrate further up the development → applied research → basic research hierarchy, owing to internal subsystem uncertainty and uncertainty in links between subsystems. At project level it is typically difficult to make sound predictions of time and cost of the innovation stage, or to foresee what precise form the new development will take. The traditional linear, rational model of innovation does not constitute a realistic or useful interpretation of what appears to be a non-linear, non-rational process; however, as yet, knowledge of the causes and consequences of technological change is still at an elementary level and adequate models to replace the rational model 'borrowed' from other areas of study have not yet been developed, (see Gold, 1971, chap. 10).

In the next chapter, attention is concentrated on specific approaches to the R & D problem, and their assumptions and method related to what has been developed in this chapter with respect to R & D projects in the firm.

3 Models of Technological Change in the Firm

The approaches discussed below illustrate in different ways some of the problems involved in formulating rational models of technological innovation. In Chapter 2 some behavioural characteristics of innovation were discussed and the importance of uncertainty emphasised; here the relevance of the different approaches is assessed in the light of the conclusions and speculations of that chapter.

The difficulties of applying rational project-based approaches are indicated in the sections below dealing with the neoclassical theory of the firm and decision-making under uncertainty. The approaches have a further link in that they have been typically applied to routine problems of resource allocation in which replicability of events is a prime assumption (though, as we shall see, Bayesian analysis is an exception in this respect). However, in both cases it has been suggested that they can be applied successfully to analysis of allocations for R & D. Since research and development is an activity which typically is not describable in terms of the assumptions of these approaches, it will be suggested that their application in this area is generally inappropriate.

The other two approaches discussed, the behavioural theory of the firm (Cyert and March, 1963) and the theory of the growth of the firm (Penrose, 1959), are introduced for slightly different reasons. As with the other two frameworks, difficulties of application to R & D problems are identified, but attention is also directed to concepts and interpretations which will prove of relevance in Chapter 4 (in which corporate behaviour is analysed more fully) and in Chapter 5 (in which a framework describing corporate resource allocation is developed).

In the latter part of the chapter general system theory is made use of, and the behavioural theory of the firm and the theory of the growth of the form re-examined to see to what extent they might be integrated in a general systems framework. As well as indicating areas of potential concern and application in the particular case of behavioural theory, this serves the purpose of indicating in what respect the theories may be deficient in their original formulation. Chapter 4 builds on this framework by analysing the modern corporation in terms of general system concepts and how R & D fits into corporate decision-making.

Each of the approaches will be examined in turn, particular attention being paid to the nature and implications of assumptions in each. The first to be discussed is the neoclassical theory of the firm.

THE NEOCLASSICAL THEORY OF THE FIRM

Attempts to analyse innovative activity in the firm within this framework have generally concentrated on the direction rather than the intensity of research and development. For example, an extensive literature has developed on the analysis of possible determinants of factor-saving innovations; Kennedy (1964) uses the concept of an innovation possibility frontier (in which research productivity is a direct function of research inputs) to analyse capital- and labour-saving inventions, while many other studies[1] have also been concerned with this problem. However, such analysis is restricted to the search for process innovations, and is therefore concerned with a restricted class of innovatory activity, since R & D activity is highly oriented towards the search for new products. Nelson et al. (1967, pp. 49–50) report a survey conducted in U.S. industry which found that about half of industrial R & D was concerned with the creation of new products, 40 per cent was concerned with improving existing product lines, and about 10 per cent was to improve processes.

The neoclassical theory of the firm assumes that the decision-maker makes price, output and factor allocation decisions with perfect knowledge and foresight as to what constitutes the relevant parameters of his cost and demand functions. Choices and allocations are made with the intention of maximising profits; the theory specifies objectively rational behaviour with respect to this goal. Considering the characteristics of R & D activity revealed by recent studies and described in the previous chapter, it is not surprising that there have been few attempts to analyse R & D activity in terms of the simple assumptions of neoclassical theory.[2] The conclusions of the previous chapter indicated that R & D is an inherently uncertain activity, the degree of uncertainty increasing as the basic research end of the R & D spectrum is approached. As Freeman (1974, p. 226) suggests, the more that innovation is differentiated from the existing experience and knowledge of the firm, the greater the level of uncertainty typically associated with the project. It appears that approximation to the condition of certain knowledge in neoclassical theory might be achieved only at the expense of trading off the definitive innovation characteristics of novelty and radicalness; in short, the less innovative a project, the more likely it is that neoclassical theory may be usefully applied.

This is supported by Mansfield, one of the most prolific and influential of economists in this area, in a discussion on the relevance of neoclassical theory to problems of technological change: 'with regard to many of the major issues concerning basic research, economics has little to say. As one moves towards the development end of the R & D spectrum, economics becomes more useful, but it still has a limited contribution to make' (1966, p. 486). McKie (1972), in reviewing the economics of technological change, puts it even more strongly: 'Simple equivalence of marginal cost and marginal social return is the test of welfare that we apply to practically every other allocation of resources, yet we are not remotely able to apply it at the present time to technological progress, which is itself an organic change in the use of resources' (p. 6).

There appears to be widespread agreement as to the nature of the problems encountered in applying neoclassical theory in this area.[3] If this represented the limits to which it has been suggested the theory can be applied, we would conclude that neoclassical theory is of limited relevance to problems of R & D, and recommend that its application be restricted to areas where its central assumptions are not so blatantly contradicted. However, some theorists have attempted to circumvent this problem by studying inventive activity[4] at a higher level of aggregation. Schmookler has conducted extensive empirical analysis using patent statistics as a measure of inventive activity. In an early work (1954) he tested the hypothesis that inventive activity is a function of measures of aggregate factor input at the level of the economy; in more recent work he has developed analysis, still using patenting as an index of inventive activity, at the lower level of aggregation of broadly defined industries (1962a,b,[5] 1966).

From his studies Schmookler concludes that the distribution and level of inventive activity is largely determined by demand conditions; 'demand induces the inventions that satisfy it' (1966, p. 184). Schmookler interprets demand in terms of expected sales quantities of a specified, as yet non-existent, good.[6] However, Rosenburg (1974) casts doubt on the applicability of this concept of latent demand, asking specifically what is meant by this definition; latent demand is difficult to define if invention to satisfy it has not been invented. There is a danger of a tautologous definition being able to explain any observed behaviour in terms of such a construct. Rosenburg identifies the need to define demand independently of observation of inventive activity. Relating Schmookler's empirical evidence back to a neoclassical micro-framework results in interpretation of analysis being vulnerable to criticism of the type referred to in the previous chapter.

In this context, Machlup (1962b) argues that at a certain level of aggregation it is meaningful to talk of a production function for inventions. Machlup bases his interpretation firmly in a rational interpretation of the inventive process, suggesting that 'The inventive process . . . has become more methodical than it used to be in earlier times, more systematic, mechanised and routinised' (p. 153).

Machlup contends there is sufficient evidence to suggest that the production of inventions can be expressed in the form of a production function, possibly even at establishment level. However, his conception of a supply curve and production function is based on the concept of a 'homogenised' invention whose purpose is to provide a construct 'for the sake of reasoning and discussion' (p. 155). In this way, his supply curve supposedly avoids the problem of heterogeneity of inventions.

Thus, 'homogenised' inventive output, plainly an inapplicable concept at project level, is seen as a useful theoretical constraint at higher levels of aggregation. In this respect, Gold (1971) puts forward relevant criticism:

> . . . aggregation . . . poses serious problems for output measurement in cases involving product heterogeneity. Even in the simple case of a plant which makes a single product in a range of sizes and models, total output can be physically aggregated only by disregarding all qualitative differences. But such a measure defies interpretation. For example, an increase in the number of units, without any indication of changes in the ratio of small to large, or of high quality to low, need not have any meaningful relationship even to the level of aggregate production activity. Nor does it support any meaningful concept of the average unit of output . . . such an average unit would not only represent some non-existent composite size and model, but would change from period to period. . . . these problems are further complicated in the case of multi-producer plants and firms. (p. 54)

Therefore, even if the problem of uncertainty can be ignored in aggregating measures of inventive output, proper and adequate interpretation of constructs and analysis of behavioural relationships is difficult, if not impossible. In a later chapter Gold directs attention specifically at such problems in decision concerned with allocation of resources to R & D:

> There are no authoritative measures of the 'productivity' or 'efficiency' of any corporation as a whole, nor of any subsector of corporate activities providing non-repetitive services. . . . Yet all of these have managed to function because input–output measurements are essentially a means of summarising the results of complex activity systems rather than the basis for understanding or managing the intricate and usually highly specialised processes involved. . . . In short . . . when we do not understand the system – as is patently true of R & D – we cannot devise strategically significant measures of its 'productivity' or 'efficiency' or determine its production function. (p. 241)

As Gold points out, useful and worthwhile interpretation of aggregated data is liable to result only if aggregation itself is based on appreciation of the processes involved in the relevant systems. The fact that the assumptions underlying aggregation in inventive activity negate what little consensus appears to have been reached by studies at micro-level, constitutes a further justification for reservations as to the legitimacy of such a blunt instrument in R & D analysis.

In fact there is possibly a further serious criticism to be made of the relevance of aggregation in this context. The idea of latent demand for inventions, tenuous at project level, becomes even more so at higher levels of aggregation. At project level, latent demand for a product or process providing a specified function may be conceivable in certain circumstances, e.g. a vaccine for rabies, or a new weapon system to counteract enemy developments; at higher levels it is a highly dubious concept. For example, at intermediate levels demand for goods can be conceptually and empirically distinguished on the basis of functional differentiation of categories – food, textiles, etc. However, within an industry grouping and at macro-level, the only distinguishing feature of inventions vis-a-vis existing products is the non-functional descriptive feature of novelty. How one may operationalise the concept of latent demand for inventions at aggregative levels in such circumstances is difficult to perceive.

The difficulty is compounded when one recognises the wide range of activities and industries at which R & D is typically aimed; Nelson et al. (1967, pp. 51–3) point out that R & D diversification is considerably higher than product diversification. They computed coefficients of specialisation and coverage for a number of two- and three-digit S.I.C.[7]

industries using N.S.F. data; coefficient of specialisation is a measure of percentage of R & D expenditures allocated by firms in an industry towards that industry's major products, while coefficient of coverage measures the percentage of R & D in each product field performed by firms in the industry which is the principal manufacturer of that product (see Table 3.1).

TABLE 3.1 Coefficients of specialisation and coverage for R & D activities for 1960

(in percentages)

Industry	*Specialisation*	*Coverage*
Aircraft and missiles	67.9	72.2
Chemicals	80.3	77.0
Electrical equipment and communication	48.7	56.9
Fabricated metal products	32.4	23.5
Food and kindred products	78.1	78.1
Machinery	51.4	50.5
Motor vehicles and other transportation equipment	58.1	87.3
Petroleum refining and extraction	52.6	93.4
Primary metals	58.8	74.3
Professional and scientific instruments	32.0	56.5
Rubber products	33.9	69.6

Source: Computed from data from National Science Foundation, *Research and Development in Industry 1960* (Government Printing Office, Washington, N.S.F. 63–7, 1963) pp. 80–1, by R. R. Nelson, M. J. Peck and E. D. Kalachek, *Technology, Economic Growth and Public Policy* (Brookings Institution, Washington, 1967) p. 51.

The median coefficients were 58 and 72 per cent respectively, while Nelson et al. point out that values for both coefficients estimated for existing products are typically in the 90 per cent range at the lower four-digit industry level of aggregation. This high level of R & D diversification makes it even more difficult to identify areas or points of reference for latent demand; the attention and direction of R & D is more easily switched to areas of potential interest than is the production of existing products, and consequently the area to which the concept of latent demand for invention is applicable would generally have to be much wider than the scope of the industry.

Therefore, consideration of R & D activity at project level in the neoclassical frame of reference falls foul of the criticism of rational models and analysis of R & D characteristics developed and presented earlier. Aggregation of data does not appear to offer any solution either; the rational model is still implicit, while aggregation creates significant, unavoidable problems not present at project level. As Grabowski and Mueller (1970, pp. 100–1) point out, microeconomic theory is basically a theory of a one-product firm, and as such is a highly specific and limited approach, especially if aggregation over many diverse products and firms is to be carried out. Research and development presents formidable problems for neoclassical theory in this respect and others; as Gold (1971, pp. 240–1) points out, perhaps the most surprising feature of such approaches is the apparent failure to recognise that such difficulties are pervasive and inevitable.

However, McKie (1972), while disparaging with respect to the possibility of usefully applying neoclassical micro-theory to R & D activity, suggests: 'Uncertainty has been an impenetratable barrier to *ex ante* evaluation of invention and innovation. The noteworthy progress in the theory of risk and uncertainty needs to be applied more searchingly to the economics of innovation to see whether it will help break the uncertainty barrier' (pp. 6–7 n).

In the next section we shall consider the possibility of utilising stochastic approaches to technological innovation, in particular the relevance of the work of Kenneth Arrow.

RESEARCH AND DEVELOPMENT AS UNCERTAIN ACTIVITY

Risk and uncertainty has been introduced into a number of analytical approaches to the R & D problem, for example information theory, Bayesian and non-Bayesian statistical decision-making, and operational research. Each approach differs in its assumptions and its specific area of concern. However, it is possible to a large extent to identify common ground among these approaches with respect to their analysis and interpretation of R & D activity. Substantial disagreement does exist in one particular area referred to at the end of this section.

Kenneth Arrow[8] has developed an integrated framework specifically designed for dealing with decision-making and resource allocation in this area. According to Arrow, 'Technological progress is in the first instance the reduction in uncertainty. The product of a research and development effort is an observation on the world which reduces its possible range of variation. . . . Research and development is . . . intimately connected with the problems of uncertainty *reduction*[9] which have been the objects of research in mathematical statistics and information theory' (1971a, pp. 166–7).[10]

Arrow defines uncertainty as implying that the decision-making agent does not know the state of the world, state of the world being a description so complete that, if true and known, the consequences of every action would be known (1971b, p. 45). An *event* is a set of states of the world which satisfy some given condition (1971b, p. 46); invention is interpretable as event.

Elsewhere, (Arrow, 1962), invention is defined as the production of new knowledge; interpretation of technical progress and invention as both reduction in uncertainty and production of new information is consistent with information theory which interprets information as the negation of uncertainty (see Marschak, 1968).

However, such analysis is essentially a modified version of the rational model. Arrow suggests (1971a, pp. 168–9) that while *a priori* probabilities of success or failure of specific activities may be small (activity being interpretable as a project or section of project), *a posteriori* probabilities of successful projects are 1 by definition. *A posteriori* probability can then be used in calculating probabilities of successful repetition of stages in R & D activity: 'At each stage, then, something is learned with regard to the probability distribution of outcomes for future repetitions of the activity, . . . definite methods of computing the optimal solutions exist' (1971a, p. 169).

While it would be redundant to reproduce all the relevant criticism of the rational model cited earlier, Gold again makes specific and pertinent criticism in this context: 'scientists and engineers do not have access to widely accepted models of the terrain beyond current research frontiers (including the identification of promising targets and of the means as well as the risks of reaching them)' (1971, p. 229).

Gold cites evidence, some of which has already been referred to in criticism of the rational model in the previous chapter (e.g. Jewkes et al., 1969; Klein, 1962, pp. 480, 508; Marshall and Meckling, 1962, p. 463). In such circumstances it is legitimate to question in what meaningful sense technical progress can be said to reduce uncertainty. An R & D project might be stimulated by unanticipated indications of possibilities in a particular area, undergo research on and development of particular ideas, and finally abort or innovate according to the subsequent failure or success of the project. However, since the project has moved from a situation in which the idea was not even conceived through intermediate stages to one in which it has actually been introduced, it is questionable that this is interpretable as uncertainty reduction; originally the invention, perceivable in intermediate stages as being uncertain, was not even recognised as a possibility. Yet if we accept Gold's claim above, such a situation is typical rather than exceptional.

Interpreting R & D activity as uncertainty reduction implies that ignorance is uncertainty;[11] as has been pointed out in a related context, uncertainty is not simply ignorance but involves knowledge of possible future states (Loasby, 1967, p. 305). Specification of uncertainty is with respect to possible events and states of the world, and that this is implicit in Arrow's approach is affirmed in his statement that R & D reduces the possible range of variation of the world.[12] Only if the decision-maker is working towards selection of the 'true' state of the world from an array of recognised alternatives can increase in knowledge be equated with uncertainty reduction; but, as Gold points out above, the typical circumstances under which R & D decision-makers operate is one of ignorance, not uncertainty.

In fact, Mulkay (1972), interpreting scientific discovery as innovation in a study of the sociology of science, states: 'many innovations take the form of a discovery of a new area of ignorance which has not previously been defined at all' (p. 39), and Bradbury (1974) comments:

> It is usual to characterise R & D as an uncertainty reduction procedure. This I believe to be a fair characterisation of much of industrial R & D, but unless one is prepared to regard ignorance as infinite uncertainty, the characterisation is a poor fit to the exploratory and search part of R & D. Search can rudely disturb the bliss of ignorance by creating a great deal of uncertainty which it is the role of subsequent product and process development to reduce. (p. 394).

Boulding (1956), in developing the concept of 'image', or subjective knowledge governing behaviour, analyses the possible effects information may have on a cognitive 'image' of the world; the image may be unaffected, qualitatively changed, clarified, *or* information may add to doubt or uncertainty in the image (pp. 7–10). As each of these outlines indicates, knowledge is not the simple inverse of uncertainty.

Thus, the stochastic approach ignores an essential and complementary characteristic of R & D to uncertainty reduction, that of uncertainty *generation* – finding new problems and possible inventions for future exploitation, uncovering and developing new areas of uncertainty. While R & D may be characterised as the development of new knowledge, this is by no means the same as uncertainty reduction, and in some types of

R & D work is its antithesis. Thus, Machlup, while developing a different version of the rational model discussed in the previous section (1962b), recognises that R & D is concerned both with generating and eliminating possible events: 'an invention may fulfil a task and at the same time create more tasks. . . . Fundamental discoveries and basic inventions, by definition, open up new vistas and create new opportunities for further invention' (p. 161).

Machlup distinguishes between agenda-increasing and agenda-reducing inventions; activity by subsystems is also interpretable in this respect, basic research being generally agenda-increasing (uncertainty-generating), development by implication being typically agenda- and uncertainty-reducing. Consequently it is in this latter area that rational analysis has tended to concentrate: Marschak et al. (1967) introduce their microeconomic analysis of R & D under uncertainty by immediately specifying that their study is restricted to the uncertainty-reducing development subsystem. It is of course feasible in such approaches that new opportunities, new areas of possible exploration, are suggested in the course of R & D work directed towards uncertainty reduction. However, uncertainty creation is incidental to the main purpose of R & D in such interpretation; the main responsibility and purpose of R & D is still with uncertainty reduction.

The measurement of uncertainty in approaches such as Arrow's is a central issue, and has been the subject of considerable debate. If earlier criticism of the rational model is valid, then in terms of the concepts of risk and uncertainty developed earlier, uncertainty in such approaches would of necessity refer to true or unmeasurable uncertainty, rather than measurable uncertainty or risk. Use of risk is effectively invalidated by the unique non-repetitive nature of invention and innovation.[13] However, if true uncertainty is explicitly incorporated in the decision-making framework, the rational process of decision breaks down since unambiguous maximisation of expected values is no longer possible on the basis of available information.[14]

Bayesian analysis has been developed to attempt to provide a subjectively rational framework for decision-making in this context. While it may be used in objectively rational analysis of routine situations (see Winsten, 1968, pp. 123–4), it is also used in situations such as research and development where 'true' uncertainty is present. Prior probabilities are assigned by the decision-maker to possible states of the world on the basis of subjected evaluation by the decision-maker, and choice of action is dictated by alternatives suggested by these 'best estimates'. Consequently, since it constitutes a means whereby quantitative rational analysis can be conducted, a number of theorists, including Arrow, recommend this approach as a framework for analysis of R & D.[15] It has in fact been recently suggested that once this step is recognised as 'inevitable, or expedient' it no longer serves any useful purpose to distinguish between risk and uncertainty (Borch, 1968, p. xiii).

Bayesian analysis, like the other approaches discussed in this section, regards R & D as being concerned with decision-making under uncertainty. In this respect it is subject to the same criticism of misrepresentation of R & D; generation of knowledge may involve uncertainty reduction and/or uncertainty creation. For this reason also, it is typically advocated as being of most relevance to the development (or predominantly uncertainty-reducing) end of the R & D spectrum. As such it has been applied as a partial analysis of R & D activity, and at best must remain so.

As to the potential application of Bayesian analysis in determination of R & D allocations in the aggregate, this does not avoid the problems encountered at project level; indeed, the most appropriate application of Bayesian analysis, if any, is liable to be selection of development projects *within* an overall R & D budget constraint. Bayesian analysis has been primarily suggested as an approach concerned with selection of projects, not determination of overall R & D budgets. One obvious objection to its use as a guideline for aggregate allocations to R & D is its partial description of the R & D process mentioned above, which would tend to understate the value of R & D to the firm. As to the legitimacy of aggregation within the scope of development work, this is a rather broader question outside the scope of the present analysis. One point which must be emphasised is that quantification in the face of true uncertainty places a heavy burden on individual decision-makers; the only effective test of reliability is *ex post* analysis, and of course in such circumstances there may be substantial variability in estimates.

A further point to be borne in mind is that it does not appear that firms determine R & D budgets this way. If resource allocation was totally decided at a project/product level, then the R & D budget would be decided *simultaneously* with the selection of projects. It will be suggested in the next chapter that gross allocations to the R & D function typically precede selection of projects in large firms. Even if selection of projects within the R & D budget is carried out using Bayesian technique, there appears to be some other basis on which overall budget is decided as a consequence of hierarchically organised decision-making within the firm.

R & D AS RESIDUAL: PENROSE AND THE BEHAVIOURAL THEORIES

The theories discussed here tend to characterise technological innovation as a process utilising resources that are in some sense excess or surplus to managerial requirements (see Starbuck, 1965, pp. 74–9). The main approaches referred to below are those of Cyert and March (1963) and Penrose (1959). While the previously described approaches had the project as their conceptual base, the theories discussed below constitute a radical reorientation of the theoretical focus of economic analysis in this area through their emphasis on allocation of resources rather than allocation to projects.

(*a*) *The behavioural approach*

The behavioural theory of the firm is a generic term describing the work of Cyert and March (1963) and derived or related works. These in turn base a great deal of the development of their argument on previously developed concepts, in particular the concepts of 'level of aspiration' and 'satisficing' first introduced into economic theory by H. A. Simon (1955). It is generally argued that once the limited decision-making capacity of the firm is recognised, the uncertainty and complexity of business environment invalidates the neoclassical view of the firm as an omniscient system capable of objective rationality (Cyert and March, 1963, pp. 99–101). Instead the firm learns the decision rules, actions and allocations which have previously contributed to the goals of the firm, and in this manner adapts to its environment. Instead of maximising specific decision variables, the concept of satisficing dictates search activity and the availability of alternative solutions to problems. The 'aspiration level' is the datum switching search activity on and off, itself being dictated by circumstances and past performance; increasing rates of achievement tend to pull aspiration levels upwards, though at a slower rate, while falls in performance tend to result in aspiration levels exceeding achievement (Cyert and March, 1963, p. 34).

Search activity is stimulated by a discrepancy between achievement and aspiration. If aspiration exceeds achievement, search for a solution takes place until a satisfactory solution is obtained, cost of search not entering into decision-making (Cyert and March, 1963, pp. 120–2). Consistent with this concept of 'problemistic search', Cyert and March develop a theory of innovation in which the effective motivating mechanism is that of current failure to achieve target levels (pp. 278–9); failure generates problemistic search which in turn produces solutions, resulting in innovative activity. We would therefore expect a pouring of resources into innovative search activity in conditions of crisis, while in conditions where aspiration levels are being achieved such activity would not be sponsored by the firm.

However, Cyert and March recognise that such an extension of the theory does not fit observed behaviour well, citing Mansfield (1961) as evidence. Firms do innovate, and innovate extensively, when they are successful, and the concept of problemistic search is unable to explain this. They therefore also utilise the concept of organisational slack in their analysis of innovation.

Organisational slack permits the allocation of resources to subunits, and the development of projects that have strong subunit support. Success allows the firm to sanction projects that would not be tolerated in conditions of scarcity. The motivating mechanism is the contribution that such projects make to subunit goals, and the distribution of slack is dependent on the strength of the subunit demands and influence.[16] The generation or discovery of important technological innovations is in this theory not an organisational goal, nor even a means to achieving organisational goals, but instead a consequence of subunit side-payments.

Cyert and March do not include any form of technological change as an organisational goal,[17] and indeed it would be difficult to do so since their model explicitly recognises goals capable of formulation in short-run terms, such as profit and sales. The concept of satisficing does not recognise the possibility of planning or anticipating future events or, perhaps more importantly, allocating resources to satisfy anything other than immediate needs. Attention is focused either on search for solutions to current problems (problemistic search), or in satisfying immediate subunit demands in the absence of problems (allocation of organisation slack). Strategy[18] and planning, and subsystems requiring such decisions, have no place in such analysis; indeed, since these are associated features of the construct of the objectively rational firm, their omission may be regarded as a creditable development of behavioural theory. However, in dealing with subsystems explicitly concerned with other than operating decisions,[19] emphasis on short-run influence may effectively distort interpretation of such subsystems to the extent that their fundamental nature and purpose is obscured. This is demonstrated by behavioural theory of innovation, in particular Cyert and March's interpretation, and K. E. Knight's model (1967).

Knight develops Cyert and March's theory of innovation, characterising distress innovation as typically cost-reducing, such as cutting back organisational employment, while slack innovation is associated with the search for technological innovations. Knight's model might be characterised as extreme behaviouralism: the firm is either in conditions of slack or distress, and typically exhibits one of the two polar patterns of behaviour associated with the respective states.

As Cyert and March point out (p. 279): 'Problem-orientated innovation will tend to be justifiable in the short run and directly linked to the problem. Slack innovation will tend to be difficult to justify in the short run and remotely related to any major organisational problem.' Knight asserts that, far from being able to afford the resources to search for technological innovations, the firm cannot even afford the risk and high cost of merely introducing product and process innovations (p. 485). Research and development is therefore justifiable only in slack conditions.

Slack is, however, not just a facilitator of R & D in behavioural theory, but a necessary condition for its existence. Since Cyert and March's model is a short-run adaptive model, R & D cannot contribute to organisational goals, only to subunit goals. All R & D resources may be defined as slack resources (except possibly for minor development work).[20] The corollary is that in conditions of distress R & D would tend to be eliminated, implying that R & D is an erratic volatile function with no inherent stability.[21] This is contradicted by Cyert and March themselves; they quote unpublished work by Seeber to suggest that in R & D subsystems in organisations there exist 'considerable attempts to smooth allocations so that they vary less from year to year than do revenues' (p. 274). The R & D subsystem acts with reference to standard operating procedures and identifiable aspirations.

Two studies of R & D performing firms in distress conditions, by Williamson (1964, pp. 94–9) and Schott and Grebner (1974), are of relevance here. In Williamson's case, a specialised company encountered declining and fluctuating rate of increase of sales, and current rate of sales was roughly halved from the normal rate of previous years. During a cost-reduction programme it was found that R & D in the firm over the previous decade had contributed hardly anything to profits (p. 95). Yet instead of abandoning its R & D effort, the firm reduced its staff of 165 to 52 and redirected much of its work to commercial R & D companies. Schott and Grebner (1974) found that the unchanging product policy conducted by the camera manufacturer Rollei-Werke Franke und Heidecke after the introduction of the innovative twin-lens reflex camera Rolleiflex in 1928 eventually led to stagnating sales and financial losses in the post-war period due to the development of Japanese imitators and advances in miniaturisation. After 1964, high levels of regular investment in R & D led to the introduction of substantially differentiated new products and reversed the decline in Rollei-Werke sales, just as the firm was on the edge of bankruptcy.

In neither case was R & D treated as a short-term distress solution or as organisational slack. Both firms were going through periods of financial adversity encouraging revision of priorities and squeezing of any slack which may have developed. Both firms critically evaluated and revised this operation with future survival and development in mind, and in their different ways emphasised the value placed on regular expenditure on R & D as a means of ensuring organisational viability. The chemical firm's preference for maintaining regularised allocations to R & D was paralleled by Rollei-Werke after the latter's evaluation of desired allocations to functions.

Yet there is an even more fundamental objection to the treatment of technological change in behavioural theory as an incidental phenomenon. The behavioural motivating mechanism for R & D allocations in conditions of slack (subunit side payments) presupposes the existence of an R & D subsystem. Resources are only allocated to the subunit because it already exists and demands side-payments. Yet since R & D does not contribute to short-run organisational goals, there is no reason why R & D should be started in the first place. Postulating a 'technological change' goal is not an adequate solution; the whole framework of behavioural theory is based on satisficing, short-run reactive responses to environment. Resource allocations with strategic intent, of which R & D is a subset, are designed to generate and create opportunities and actively affect the firm's environment;[22] the reactive nature of satisficing emphasises the opposite consideration, that of environment on the firm's actions in the short run.

For these reasons, Cyert and March's behavioural theory[23] has difficulty in dealing with and interpreting technological change. However, the work introduces and attempts to synthesise important concepts with respect to the firm as a resource-allocating and administrative unit. Further reference will be made to this important work, and use made of its findings and theoretical concepts.

(b) *Penrose's theory of the growth of the firm*

In a separate development of a theory of the firm, Penrose (1959) develops an approach which has similarities to behavioural theory, but is crucially different in a number of important respects. Penrose takes the point of view that growth results from the pursuit of economies and opportunities which 'disappear' once the expansion has been achieved. New opportunities for growth may arise once the expansion has been completed, perpetuating the growth process. According to Penrose:

> Economies of growth are the internal economies available to an individual firm which make expansion profitable in particular directions. They are derived from the unique collection of productive services available to it, and create for that firm a differential advantage over other firms in putting on the market new products or increased quantities of old products. At any time the availability of such economies is the result of the process . . . by which unused productive services are continually created within the firm. (p. 99)

Penrose differentiates between resources and services obtainable from these resources: 'Resources . . . include the physical things a firm buys, leases, or produces for its own use, and the people hired on terms that make them effectively part of the firm. A resource then can be viewed as a bundle of possible services' (p. 67). Services, not resources, are inputs in the production process, and resources can usually be defined independently of their use (p. 24). Penrose avoids the use of the term 'factor of production' since she suggests that it is used interchangeably to mean both resources and services in neoclassical theory.[24]

Penrose provides a mechanism for generating growth: 'The services available for expansion are the difference between the total services available to the firm and those required to operate it at the level of activity appropriate to its existing circumstances' (p. 201). Unused or residual managerial resources therefore constitute the means by which opportunities for growth are exploited since they provide excess managerial services, excess being defined as the amount over and above that required to manage the firm in its current operations. Growth is generated particularly by underemployment of managers, who seek means and opportunities to work to their full capacity (pp. 49–50). The release of unused resources is facilitated over time by 'learning by doing'.

In the development of Penrose's theoretical approach, the amount of resources required, and allocated to, current operations is invariable: 'For any given scale of operations a firm must possess resources from which it can obtain the productive services appropriate to the amounts and types of products it intends to produce' (p. 67). It is not a variable which can be subjected to discretionary variation with respect to possible substitution of resources for growth.

It might appear that R & D would be automatically interpretable as growth activity rather than operating activity in Penrose's theory. In fact, as well as potentially providing a foundation which may give the firm an advantage in some new areas, Penrose suggests that research may be useful in the production of a firm's existing products (presumably as far as development of new processes and technology is concerned) (pp. 114–15). There appears some possible ambiguity here, but it does seem that at least some R & D is interpretable by Penrose as *operating* activity oriented towards existing production, as well as having a prime role in the diversification process itself (see pp. 112–16).

While the role of R & D may be subject to interpretation according to circumstances in Penrose's theory, at a higher level of abstraction it is possible to identify correspondence between it and behavioural theory; both conceive of *residual* resources (organisational slack in behavioural theory) as being the means whereby growth is fuelled, thus implying an 'operating over expansion' priority rule. Operating problems are attended to first, then growth opportunities.

Bearing in mind the qualification that at least some R & D may be classified as operating activity in Penrose's approach, to the extent that R & D is oriented towards diversifying and extending the firm's range of activities in pursuit of growth opportunities, we would expect resource allocation to it to be unstable and erratic as in the behavioural theory of the firm. As with the latter approach, the generally stable nature of allocations to R & D,[25] and in particular the institutionalised nature of the function in large firms, runs counter to what would be expected given the broad theoretical implications of the model.

It must be emphasised that neither the behavioural theory of the firm nor the theory of the growth of the firm were developed to deal specifically with problems of resource allocation to technological change. Yet, as has been indicated above, in both approaches recognition of the importance of technological change is expressed, and both Cyert and March, and Penrose, attempt to account for resource allocation to innovative activity within their respective theoretical frameworks.

It is not intended here to provide any general attempt to refute or negate the theoretical approaches which have been discussed. Instead it is hoped that the discussion of the assumptions of each will have provided a perspective for discussion in the next chapter and the latter part of this chapter as to required developments for analysis of the problems of technological change. Tentatively, however, it would seem that in their received formulations none of the approaches discussed appears to offer a satisfactory basis for such analysis.

In the next section, general system theory is introduced. As mentioned at the start of this chapter, this provides both a means of placing behavioural theory and theory of the growth of the firm in context, as well as a framework for the analysis of the corporation in Chapter 4. The integration of both previously discussed approaches into a systems perspective is facilitated by their recognition and use of the concept of 'resource'. Unlike the neoclassical and stochastic approaches in which the emphasis is on project- or product-specific *services*, the existence of stable and enduring subunits, or other 'bundles' of resources, is a feature of both behavioural theory and theory of the growth of the firm. The requirements for extending such interpretation into a fully *open* system framework for analysis of corporate behaviour will be the concern of the next two chapters. However, we shall start by examining some of the characteristics and implications of general system theory.

GENERAL SYSTEM THEORY CLASSIFICATION OF MODELS AND SYSTEMS

As mentioned in Chapter 1, general system theory is to some extent a misnomer. It is not a testable theory or set of hypotheses as such, but is instead a framework or 'meta-theory' within which can be set lower-level hypotheses.

Its potential usefulness lies partly in the frame of reference imposed by its conceptual language. The structures and relations identified will hopefully permit an improved picture of the behaviour of the corporation to be drawn. Firstly, we shall develop a brief classification of systems in the light of this approach before reappraising behavioural theory and the theory of the growth of the firm.

General system theory is concerned with 'general aspects, correspondences and isomorphisms common to "systems"' (Von Bertalanffy, 1973, p. xviii). Systems consist of 'sets of elements standing in interrelation' (ibid., p. 37); a corporation may be regarded as a system in its own right, constituent elements (such as individuals, departments, materials and building, etc.) contributing to the structure (or 'order of parts': ibid., p. 25) and functions ('order of processes': ibid., p. 25) of the overall system. A fundamental precept of general system theory is that of hierarchic order (ibid., p. 25); for example, systems themselves may be organised into hierarchies; corporations may be regarded as elements in an industry 'system' while R & D, marketing, production, etc., are interpretable as systems in their own right, or subsystems in the corporation 'system'.

As well as systems generally being capable of arrangement in hierarchies, so too are models of systems. Boulding (1968) provided 'a possible arrangement of "levels" of theoretical discourse' (p. 6) which are reproduced and further interpreted by Von Bertalanffy (see Table 3.2). Boulding identifies nine levels of systems and associated models, but three levels are of particular interest to the present analysis, those at the second, third and fourth levels respectively.

TABLE 3.2 An informal survey of main levels in the hierarchy of systems (partly in pursuance of Boulding, 1968)

Level	*Description and examples*	*Theory and models*
1. Static structures	Atoms, molecules, crystals, biological structures from the electron-microscopic to the macroscopic level	e.g. structural formulas of chemistry; crystallography; anatomical descriptions
2. Clock works	Clocks, conventional machines in general solar systems	Conventional physics, such as laws of mechanics (Newton, Einstein and others)
3. Control mechanisms	Thermostat, servomechanisms, homeostatic mechanism in organisms	Cybernetics; feedback and information theory
4. Open systems	Flame, cells and organisms in general	(a) Expansion of physical theory to systems maintaining themselves in flow of matter (metabolism) (b) Information storage in genetic code (DNA) Connection of (a) and (b) presently unclear
5. Lower organisms	'Plant-like' organisms: increasing differentiation of system (so-called 'division of labour' in the organism); distinction of reproduction and functional individual ('germ track and soma')	

TABLE 3.2 (*contd.*)

Level	*Description and examples*	*Theory and models*
6. Animals	Increasing importance of traffic in information (evolution of receptors, nervous systems); learning; beginnings of consciousness	Beginnings in automata theory (S–R relations), feedback (regulatory phenomena), autonomous behaviour (relaxation oscillations), etc.
7. Man	Symbolism; past and future, self and world, self-awareness, etc., as consequences; communication by language, etc.	Incipient theory of symbolism
8. Socio-cultural systems	Populations of organisms (human included); symbol-determined communities (cultures) in man only	Statistical and dynamic laws in population dynamics, economics, possibly history; beginnings of theory of cultural systems
9. Symbolic systems	Language, logic, mathematics, sciences, arts, morals, etc.	Algorithms of symbols (e.g. mathematics, grammar); 'rules of the game' such as in visual arts, music, etc.

Note: This survey is impressionistic and intuitive with no claim for logical rigour. Higher levels as a rule presuppose lower ones (e.g. life phenomena those at the physico-chemical level, socio-cultural phenomena the level of human activity, etc.); but the relation of levels requires clarification in each case (cf. problems such as open system and genetic code as apparent prerequisites of 'life'; relation of 'conceptual' to 'real' systems, etc.). In this sense, the survey suggests both the limits of reductionism and the gaps in actual knowledge.

Source: L. von Bertalanffy, *General System Theory* (Penguin Books, London, 1973) pp. 26–7. Reprinted by permission of Penguin Books Ltd.

Second-level systems are simple systems with 'predetermined necessary motions' (Boulding, 1968, p. 6). In such systems the final state is determined by the initial conditions of the system (Von Bertalanffy, 1973, p. 39); the solar system itself is an example of such a system, in which the behaviour of the system as a whole may be effectively studied without reference to other systems. Such systems are interpretable as closed systems, 'systems . . . considered to be isolated from their environment' (Von Bertalanffy, 1973, p. 38). Such systems do not interact with their environment, and consequently their system boundary may be regarded as the limit of relationships of relevance to system behaviour. There is no feedback of stumuli or response from system to environment or vice versa (see Fig. 3.1).

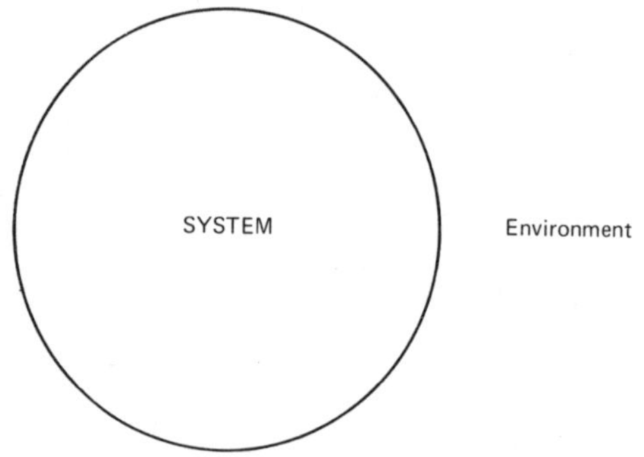

Fig. 3.1. Closed system

Third-level systems are identified as cybernetic systems; this is the level of the control mechanism (Boulding, 1968, p. 7). Homeostatic mechanisms are integral features of such systems, homeostasis being the maintenance of the internal state of the system by regulating mechanisms which correct deviations from this internal state (see Von Bertalanffy, 1968, p. 17–18). Such systems operate by means of stimuli from the environment, homeostasis being achieved by means of an iterative feedback process (see Fig. 3.2).

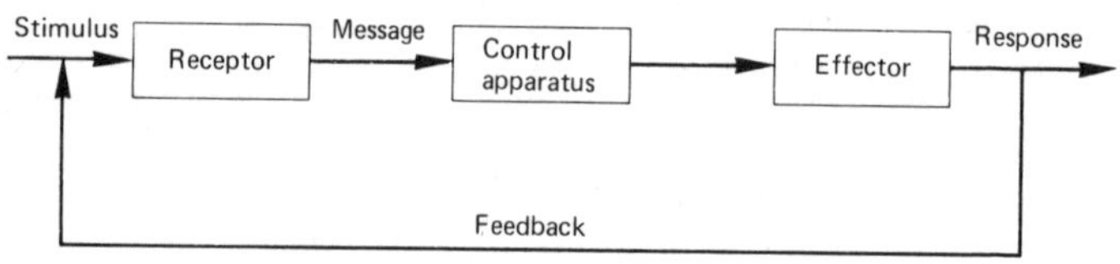

Fig. 3.2. Simple feedback scheme

Source: L. Von Bertalanffy, *General System Theory*, Penguin University Books, London, 1973, p. 42.

The response of a system to deviations in its environment is determined by the structural arrangement of the regulatory mechanism itself. Possible responses and actions of the system are therefore preprogrammed and limited (Von Bertalanffy, 1973, p. 43). At the third level there is no autonomous action which the system is recognised as being capable of. Consequently, stimuli[26] may be represented as being unidirectional, from environment to system, as in Fig. 3.3.

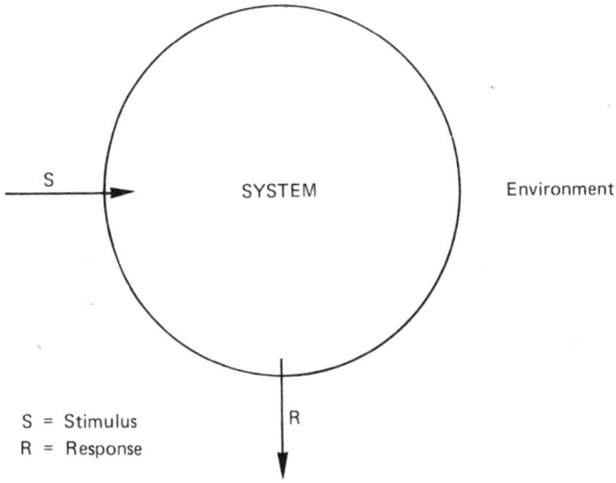

FIG. 3.3. Cybernetic system

The fourth level is that of the 'open system' or 'self-maintaining structure' (Boulding, 1968, p. 7). There are a number of characteristics of open systems which will be of relevance in subsequent analysis,[27] but there is in particular a necessary distinction between cybernetic and open systems crucial to the present discussion; this is made by Von Bertalanffy (1968, p. 19) who states that 'spontaneous mass activities' of the system in 'dynamic interaction' with its environment characterise the open system, in contradistinction to the characterisation of the cybernetic system as a 'reactive system' answering outside stimuli according to programmed responses. Emery and Trist (1969, p. 282) adapt the concept of open system to organisational systems, a defining characteristic of such systems being that they 'may spontaneously reorganise towards states of greater heterogeneity and complexity'.

The open system is thus capable of autonomous regulation and action (see Fig. 3.4).

These distinctions constitute probably the most elementary basis possible for a taxonomy of systems at the second, third and fourth levels, yet they provide fundamental behavioural characteristics on which complex theories are developed.[28] They are arranged hierarchically in order of complexity of possible relationships and behaviour patterns, the hierarchic ordering facilitating analysis of types of system behaviour. However, as Von Bertalanffy points out, the prior existence of open system may be a necessary precondition of third-level homeostatic mechanisms in living systems: 'Reflex reactions answering external

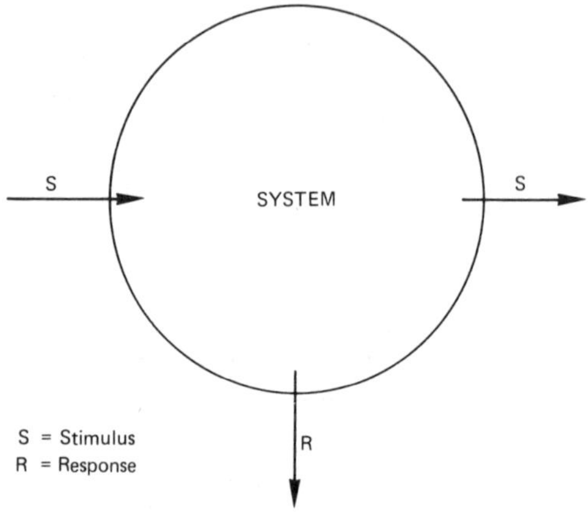

FIG. 3.4. Open system

stimuli and following a structured path appear to be superimposed . . . as secondary regulatory mechanisms. . . . In this sense it appears that in development and evolution . . . open system precedes mechanisation (structured arrangements particularly of a feedback nature)' (1968, p. 19).

Extensive and elaborate homeostatic control mechanisms may therefore be a feature of mature and developed open systems.[29] This would be anticipated from the hierarchical arrangement in Boulding's taxonomy; he points out that higher levels may be regarded as incorporating all levels below, a corollary being that useful insights may be gained by partial analysis of higher-level systems using lower-level models.

Even these basic system definitions are sufficient to provide some useful interpretation of the theories discussed in the earlier sections. For example, in discussions of third-level systems Boulding (1968) associates this level with the thermostat:

> This differs from the stable equilibrium system mainly in the fact that the transmission and interpretation of information is an essential part of the system. As a result of this, the equilibrium position is not merely determined by the equations of the system, but the system . . . will maintain *any* temperature at which it can be set, . . . the essential variable . . . is the *difference* between an 'observed' or 'recorded' value of the maintained variable and its 'ideal' value. If this difference is not zero the system moves so as to diminish it; thus the furnace sends up heat when . . . 'too cold' and is turned off when the recorded temperature is 'too hot'. (p. 7)

There are close similarities between the thermostat system and the satisficing mechanism which constitutes the basis of the behavioural theory of the firm. Action is triggered by failure to achieve the 'ideal value' in the case of the thermostat, and the aspiration level in the case of behavioural theory. The 'ideal value' and level of aspiration are not determined solely by the characteristics and specifications of the thermostat and satisficing mechanism respectively, but can move to a range of values within limits, depending on circumstances. In the case of satisficing, the current aspiration level is determined by historic levels of aspiration and achievement, and relationships between these variables. Both systems are reactive systems, requiring a stimulus signifying deviation from a particular level in a specified direction before action is taken. If temperature is below 'ideal value' the furnace operates; if achievement is below aspiration, problemistic search is undertaken. Attainment of ideal temperature and solution to problems terminates furnace operations and problemistic search respectively.

However, problemistic search, the system response in behavioural theory, is significantly different from normal homeostatic behaviour. Response in the thermostat results in output into the environment of a homogeneous nature according to pre-programmed rules; typically, problemistic search generates differentiated alternative solutions and since the solution may not be previously known to the firm, programming is necessarily incomplete.[30] Despite this, there is substantial routinisation of problemistic search (Cyert and March, 1963, p. 80). According to Cyert and March, 'rules for search . . . reflect simple concepts of causality . . . search is based initially on two simple rules:

(i) search in the neighbourhood of the problem symptom and
(ii) search in the neighbourhood of the current alternative (1963, p. 121).

The organisation learns preferred manners and directions of search, and only invokes more complex, distant or unfamiliar search rules or patterns if the routinised search procedure is unsuccessful (Cyert and March, 1963, pp. 122, 124).

Therefore, satisficing would appear to operate in a manner appropriate to a third-level system in the Boulding taxonomy. This reflects Cyert and March's emphasis on the importance of short-run adaptation: 'so long as the environment of the firm is unstable (and unpredictably unstable, the heart of the theory must be the process of short-run adaptive reactions' (1963, p. 100). Since many standard operating procedures[31] are stable in the short run and direct and influence activities and decisions, they parallel the role of structural arrangements in homeostatic systems. The behavioural theory may be interpreted as essentially based on third-level models of corporate behaviour, albeit of a specialised and sophisticated nature.

This permits useful inferences to be drawn as to the contribution and possible defects of behavioural theory. If we accept Von Bertalanffy's claim that open system generally precedes the development of homeostatic mechanisms, then the behavioural theory concentrates on only one aspect of open system behaviour, that of secondary regulatory mechanisms utilising fixed structures and responses. Yet if these are generally only characteristics of open systems, and open system is a necessary condition for development of satisficing mechanisms, we should expect fourth-level models to be necessary for a comprehensive and integrated development of behavioural theory; otherwise behavioural theory remains a partial and unsound framework. As Marris (1963) comments with respect to the earlier satisficing models of March and Simon (1958), 'The lobster-like behaviour of a "March and Simon" model might well be observed in the lower levels of business organisations, where the decision rules are necessarily rigid, but it is more difficult to imagine in a board of directors' (p. 207). This does not necessarily imply replacing any of the concepts of behavioural theory, since, as Boulding points out, lower-level systems may be incorporated in higher levels, and by inference lower-level models may be incorporated in higher models. This prompts two broad questions: firstly, what additional organisation operations and decisions would be expected to be associated with a move from a third- to a fourth-level behavioural model; and secondly, how significant are these operations and decisions for analysis and interpretation of technological change and R & D in the firm?

Both questions will be discussed in the next chapter. The problem of what type of organisational behaviour might be of concern to an open system model is in fact a prime concern of Ansoff (1965a, 1969). He identifies the concern of Cyert and March with repetitive operating decisions and suggests that strategic decisions, in which the firm autonomously organises, allocates resources and anticipates environmental effects, are of crucial concern to a study of the behaviour of firms. Of particular interest is his development of a strategic decision method which includes in particular: monitoring the environment for changes and searching for attractive product opportunities; consideration of allocation of resources between current operations and possible future opportunities; evaluation of competitive advantage, long-term potential and possible joint effects (synergy) of opportunities; and coping with potentially antagonistic objectives (Ansoff, 1965a, p. 28).

Some of these activities may be at least partially dealt with by the behavioural theory of the firm.[32] However, the distinguishing characteristic of strategic decision is the autonomous organisation and development of new products and markets (Ansoff, 1965a, pp. 100–1); the firm itself is organising to affect and change its environment, rather than passively responding to environmental stimuli as a simple homeostatic system. The development of the concept of strategic decision may therefore be seen as contributing to a fourth-level model of organisational behaviour.

The role of strategy as a director and motivator of R & D is also emphasised by Ansoff: 'Strategic change [is] a shift in the product or service mix . . . and/or markets. . . . A key step in the shift is the discovery of a product-market idea. . . . *Discovery* of a novel product-market idea is a creative act; the idea must be either invented inside the firm (usually by R & D or marketing departments), or searched out from among opportunities' (1969, p. 21). Of particular relevance is his earlier comment on the implications of failure to develop strategy:

> In the absence of strategy, there are no rules to guide the search for new opportunities, both inside and outside the firm. Internally, the research and development department has no guidelines for its contribution to diversification. . . . Thus the firm as a whole either passively waits for opportunities, or pursues a 'buckshot' search technique. (1965a, p. 102)

This illustrates the problem of analysis of R & D as organisational slack or distress innovation in behavioural theory. Unless analysis of R & D and technological innovation is carried out in the context of open system models of the firm, R & D presents severe difficulties for interpretation, since in the absence of strategy it tends to be undirected or reactive. Treatment of R & D as a third-level response to environmental change constitutes an unsatisfactory description of such non-programmed activity, typically characterised by long lead times. If R & D projects acted as problemistic search response to environmental problems, then even if a satisfactory answer to the original stimulus could be found, environmental change in the interim is liable to invalidate the derived solution. Innovation, with its long gestation period and original features, requires organisation and planning. Opportunities and challenges must be *anticipated*, emphasising the centrality of the open system concept of strategic formulation in this area.

As far as Penrose's analysis is concerned, general system theory may also clarify the possible limitations of the model. There is no active consideration in this approach of how much should be allocated to search for new opportunities as opposed to current operations; this is entirely dictated and determined by operating requirements. If current operations require changed levels of resources, residual unused productive services are varied appropriately.

In short, as far as the formal mechanism for allocating resources to growth opportunities is concerned, the decision is second-level. The environment does not affect the distribution of resources between growth opportunities and operations; this is entirely dictated by internal operating requirements. In this respect the firm operates as a closed system, and to the extent that R & D may be characterised as search for growth opportunities, it will have the implicit features of potential vulnerability and instability suggested earlier.

Yet Penrose does emphasise the importance of ability to anticipate and prepare for changes in environment (1959, pp. 41–2, 114), and identifies quality of management as crucial:

> . . . at the very least we have to assume that the firm is eager and willing to find opportunities and is not hindered in acting on them by 'abnormally' incompetent management. In other words the firms with which we shall be concerned are enterprising and possess competent management; our analysis of the processes, possibilities and direction of growth proceeds on the assumption that these qualities are present in the firm. (p. 32)

In effect this means that the firm is assumed to have the innate ability to perceive and generate growth opportunities through its managerial resources, in effect to act as an open system. Yet this informal description of a well-managed firm is not reflected in Penrose's formal specification of the mechanism whereby resources are allocated; the environment plays no part in this process.

This suggests an inconsistency in the model: the firm is a closed system for the purposes of allocating resources between growth and operations, yet becomes an open system once the allocation is made. This is not impossible, but it is an unconvincing formulation of the model; if management is 'eager', 'competent' and conscious of the importance of environment, we should expect allocation of resources to growth opportunities to be responsive to environmental conditions. This is essential if the firm is to be specified as a fully open system.

That the modern corporation does in fact generally operate as a fully open system is a central argument of the next chapter, and evidence is presented in support of stable managerial preferences for R & D in particular. We shall also identify the specific features of open system organisation that will be further considered in the process of developing a framework in Chapter 5. In doing so, the tentative conclusions reached so far will be related to a description of how the modern corporation is typically administered.

SUMMARY

Application of the neoclassical theory of the firm and related project-based stochastic approaches have been of limited and indeed questionable utility in the area of technological change. The highly complex and uncertain process of generating technological innovation involves concepts and techniques alien to the specialised knowledge and training of economists, as well as directly contradicting basic assumptions on which rational economic models are built.

Subsequent development of theories of the firm by Cyert and March, and Penrose, present rather different views of the firm as a construct from those of the above approaches. Aggregation of worthwhile projects is no longer automatically the means by which allocations to functions are derived. Subunits are recognised to have identity in their own right in the behavioural theory, and in Penrose's model residual managerial services provide the source of opportunities for growth. Both approaches identify systemic properties of corporate behaviour beyond that of the project, and introduce concepts which will be utilised in Chapter 5 in developing a framework of corporate behaviour.

Yet both the behavioural and Penrosian approaches appear to describe only partially the process by which corporate behaviour is determined. In particular, the deliberate and widespread practice by firms of allocating funds to the search for new strategies and technological innovations is not adequately explained by such theories. Using a general systems taxonomy facilitates explanation of the source of such neglect, the failure to treat the firm as a fully open system.

4 Industrial Research and Development at Functional Level in the Modern Corporation

We have already utilised general systems concepts to analyse some definitive features of theoretical approaches to technological change. The idea of open system was suggested as a necessary construct for analysis of strategic change and proved useful for provisional classification of behavioural and Penrosian theories. In this chapter the significance of strategic change in modern corporate behaviour is discussed in relation to environmental developments. The evolution of modern multi-divisional corporate structure is analysed partly in terms of increased need for institutionalising the formulation of strategic change, and the creation of R & D departments is seen as an example of this response. Later in the chapter, R & D subsystems are examined in the light of the subsystem hierarchy suggested in Chapter 2; the characteristics identified will prove useful in Chapter 7 when the possible determinants of basic research activity are examined.

From the arguments of the last chapter, to the extent that strategic change is an important concern of firms, the firm must be regarded as an open system and modified accordingly. Boulding's taxonomy has proved useful in providing a brief introductory description of types of systems including open systems; however, the behaviour of open systems in particular may be analysed more deeply in terms of a number of characteristics which facilitate analysis and exploration of open system behaviour. In particular, Katz and Kahn (1966, pp. 19–26) develop the work of Von Bertalanffy and from his analysis of the systems identify nine characteristics which appear to define all open systems. The first four characteristics are input, throughput, output, and the cyclic nature of events in systems. The fifth characteristic is that of negative entropy; Katz and Kahn state that 'The entropy process is a universal part of nature in which all forms of organisation move towards disorganization or death. The open system, however, by importing more energy from its environment than it expends, can store energy and acquire negative entropy' (p. 21).

These five characteristics constitute basic features of open system. However, it is with the rather more subtle behavioural implications of the four, relatively complex, remaining characteristics of negative feedback, steady state, differentiation and equifinality that open system analysis provides guidelines for study of corporate behaviour and development of potentially useful models. The applicability of each will be discussed in the context of appropriate description of corporate organisation and behaviour in this and the next chapter. We shall begin by looking at the historical development of the corporation.

THE EVOLUTION OF THE MODERN CORPORATION

During the course of the twentieth century the administration of the modern industrial firm has undergone radical change. There have been substantial quantitative changes in industrial organisation with a tendency towards increasing size of corporations[1] and also concentration of industrial activity in a diminishing number of corporations.[2] Industrial production has become dominated by numerically few corporations, and it is with respect to the conduct and behaviour of these large institutions that attention must be focused if industrial activity is to be analysed.

These changes have been accompanied by qualitative changes in corporation management. While generalisations must be heavily qualified it is possible to identify distinctive trends in the development of novel administrative structures and processes. Certain factors and influences are discernible as being particularly influential and decisive with respect to administrative evolution.

Chandler (1962)[3] describes the typical industrial enterprise in the United States before 1850 as being very small, with little need for a full-time organiser or clearly defined administrative structure. Routine activities (buying, selling, supervision of operations) tended to be their main preoccupation, long-term plans or decisions being rarely required (pp. 22–3). However, the rapid growth of the railway system in the latter half of the nineteenth century greatly expanded the potential market for goods and services and both facilitated and necessitated that firms extend and subdivide their operations. This expansion required new managerial skills of co-ordination, evaluation and planning of the specialised functions (Chandler, 1962, p. 27). By the end of the century there had developed numerous large privately owned firms typically operating in a single product or market area such as mining, meat-packing and steel. They tended to be organised on a multi-functional basis in which there was a series of specialised subunits, such as marketing, production, purchasing and sales. These private enterprises were frequently vertically integrated with divisional co-ordination directed from a central head office. The beginning of the twentieth century saw the widespread adoption of unitary, centralised, functionally departmentalised organisation (Chandler, 1962, pp. 27–50). This is defined as unitary or U-form organisation (see Fig. 4.1).

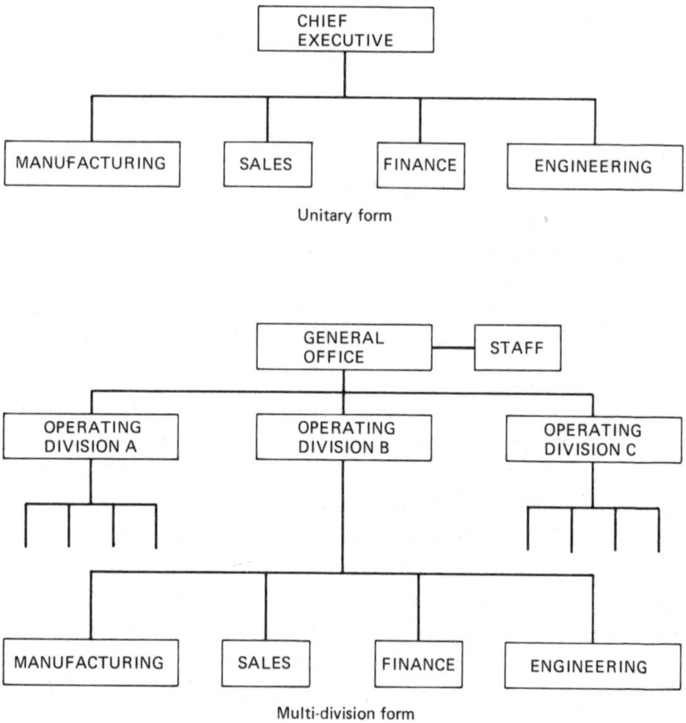

Fig. 4.1.

Source: O. E. Williamson, *Corporate Control and Business Behaviour*, p. 116 © 1970, reprinted by permission of Prentice-Hall Inc., Englewood Cliffs, N.J.

The development of large and administratively complex organisations began to increase the responsibility and power of professional management. The diffusion of ownership of corporations further removed direct control of owners over corporation management.[4] Whereas before the individual entrepreneur tended to be identified with the management of the enterprise, the dispersion of ownership through thousands of shareholders owing to the growth of firms[5] increasingly devolved power to those involved in the day-to-day running of the corporation. Growth of corporations tended to create new administrative problems requiring continuous and professional managerial skills for their solution.

However, growth could take different forms, and the form of growth dictated the degree and type of administrative problems. Growth might result from expansion of the firm's existing lines to the same type of customers, or from a search for new markets and sources of supplies, or finally from the opening of new markets through diversification. Simple expansion of existing markets brought minimal qualitative change in administrative problems, and these industries tended to be still administered through the U-form,[6] or centralised, functionally departmentalised organisational structure.[7]

This simple, extrapolative growth avoided a fundamental weakness of the U-form. A few managers were effectively responsible for the major decisions which management of the overall enterprise required. Since their experience and knowledge tended to be restricted to a single functional activity and problem area, their ability to deal with complex problems tended to be constrained to these boundaries. As long as familiar problems had to be dealt with through this simple expansion of existing products, this was not critical; however, if growth required venturing into new areas, particularly in terms of markets, then vulnerability of the centralised U-form organisational structure to the increasing complexity of decision-making required, became apparent. Diversification increased the number, complexity and unfamiliarity of both operational and strategic decision-making, making systematic co-ordination of activities an extremely complex and difficult task and creating a great strain on the limited power of existing U-form structures to cope with the situation (Chandler, 1962, pp. 50–3).

The multi-divisional or M-form organisation structure (see Fig. 4.1) was introduced to provide a decentralised solution to the increasingly complex problem of organising highly diversified firms, a number of firms instigating the new organisational structure in the 1920s in the United States. In the M-form, a general office co-ordinates, evaluates, formulates strategies and allocates resources to a number of quasi-autonomous divisions. Each division administers a major product or geographical area, with a central office administering a number of departments which operate on a functional basis similar to that of the U-form. At the lowest level, field units (branches, plants or works) are responsible to departments. General office is thus free to concentrate on broad strategic decisions of concern to the whole enterprise, while the lower levels of the organisation are more concerned with operational and minor strategic responsibilities (Chandler, 1962, pp. 11–13). General office still consists of a few crucial decision-makers who are responsible for the regulation and general direction of the enterprise; however, while in a sense still specialists dealing with a limited range of problems, their specialist responsibility and psychological commitment is with respect to the broad range of strategic, long-term plans and decisions pertaining to organisational charge and development. They are a group of 'generalists' (Chandler, 1962, p. 383) and in fact might be described as specialist generalists, freed from the routine, operating decisions required at lower levels. Bower (1970, p. 190) describes this as involving the creation of a 'new order of complexity' in the administration of the corporation, the hierarchical organisation of the corporation being further elaborated from the U-form.

The M-form development facilitated transfer of resources to profitable areas, creation and termination of divisions, and the selection of the most able managers. It also encouraged intra-firm competition for resources between divisional managers as well as market competition, since enterprise resources were allocated on the basis of measurable performance (Channon, 1973, p. 4). The M-form constituted a rationalisation of the organisation of the business firm and institutionalised active consideration of strategic change. The firm attempts to impose change on its environment rather than simply predicting and reacting to environmental change.[8] This may be interpreted in terms of attempting to establish a 'negotiated environment' (Cyert and March, 1963, pp. 119–20); 'Rather than treat the environment as exogenous and to be predicted [firms] seek ways to make it controllable' (p. 120).[9]

The large modern firm is therefore typically highly diversified and uses the M-form organisational structure.[10] Strategic decision-making is integrated as a recognised and institutionalised feature of corporate action. The firm can no longer be regarded simply as the manufacturer of a class of products, since in the M-form it becomes the vehicle for strategic change and development of innovations (Schon, 1971, pp. 63–4).

However, such developments also contributed towards qualitative changes in environment. What is systemic change in one frame of reference becomes environmental change in another. Complex and dynamic environments or 'turbulent fields' (Emery and Trist, 1965) have become features of contemporary organisational concern. Terreberry (1968) suggests that such development effectively precludes long-range planning and strategy; the environment becomes inherently uncertain when turbulent fields develop, creating great, if not insuperable, difficulties for rational decision-making and effectively inhibiting the usefulness of long-term strategy.

In those circumstances, the ability to effect long-term strategic change has been questioned extensively. March and Simon (1958, pp. 176–7) and Cyert and March (1963, esp. chap. 6) incorporate the satisficing mechanism as a heuristic means of coping with adjustment to environment, rather than planning change in complex, uncertain circumstances. Lindblom (1959) and Braybrooke and Lindblom (1963) suggest that 'disjointed incrementalism', or development of consensus policy to complex problems through a series of approximative solutions, is not only a prevalent form of decision-making but also relevant and realistic. Jantsch (1969) recommends that development of technological change should be an iterative adaptive process, in the face of uncertainty as to technological and environmental futures. Loasby (1967) points out the danger of commitment to long-term plans in conditions of rapid change in uncertain environments. All these are consistent with the rationale of Cyert and March's model described earlier, that in unstable and uncertain environments theoretical development of models must be based on short-run adaptive reactions (1963, pp. 100–1).

If strategic decision-making is effectively eliminated by turbulent environments, there would be a deep irony in the fact that a central justification for creation of M-form organisations in the first place (facilitation of strategy through separation of strategic and operating decision-making) is itself a major contributory factor to the creation of turbulent fields. Yet Ansoff (1969, p. 11) dissents from the consensus view of decision-making in turbulent environment:

> Since the early 1950s, confronted with the growing variability of the business environment, business managers have become increasingly concerned with finding rational and foresightful ways of adjusting to and exploiting environmental change. Out of this concern grew practical management approaches and systems, as well as an increasing understanding of the problem of the relationship between the firm and its environments.

Schon (1971, chap. 3) supports this view of the increasing importance of strategy and identifies this with the widespread adoption of the M-form after the Second World War (pp. 59–64). Case studies of highly diversified firms also tend to emphasise the importance of active formulation and implementation of strategy.[11]

In fact, disagreement over the contemporary relevance and effectiveness of corporate strategy is not as substantial as might first appear to be the case. There is no doubt that unstable environments increase the potential utility of anticipating environmental change and planning future actions; if environmental change is slow or infrequent, lagged reaction to environmental stimuli may be sufficient to maintain corporation viability and ensure survival. However, in dynamic and rapidly changing environments lagged reaction may not be sufficient to permit the firm to keep pace with developments; response to changing circumstances may involve a significant time-lag, especially if radical innovation is required, by which time circumstances may have changed again rendering the response redundant or inappropriate. In such circumstances, anticipatory and autonomous development of strategic action may be desirable.[12] Further, few analysts would disagree that instability and rapid change in environmental parameters create some difficulties in formulating and implementing strategic change which do not exist in stable conditions.[13]

Consequently, in times of rapid environmental change the increased need for strategy tends to be paralleled by increased difficulties in effective development and application of strategy. To a substantial extent, the controversy over whether strategy formulation or adaptive response constitutes the prevalent or appropriate corporate action revolves around disagreement over the relative significance of these issues. While the importance of such disagreement should not be understated, it may be helpful to reformulate the problem to focus on the more general consideration of how the modern corporation has evolved in terms of its relations with its environment; strategy tends to be generally interpreted as referring to *specific* problems and innovations and, as was shown in the previous chapters, the prevalence of uncertainty may mitigate against rational analysis of individual decision or plans. While decision-making for strategic purposes constitutes a very important area in analysing resource allocation for technological change, whether conducted by intuitive, rational or other means, we shall suggest that such concern may be a problem for *lower-level* analysis of corporate decision-making. We shall suggest that it is possible to interpret the corporation in terms of a hierarchy of decision-making levels, and that the problems of specific strategies may be subsumed into a lower order of concern than that of the R & D budget decision. This is the concern of the remainder of the chapter, and we shall start by suggesting how the organisation may be interpreted as a hierarchical system.

THE FIRM AND ITS ENVIRONMENT: EVOLUTION OF ORGANISATIONAL HIERARCHY

According to Simon (1965), 'Large organisations are almost universally hierarchical in structure. That is to say, they are divided into units which are subdivided into smaller units, which are, in turn, subdivided, and so on. They are also generally hierarchical in imposing on this system of successive partitionings a pyramidal authority structure' (p. 99). Thus, organisations typically exhibit the important systems characteristic of hierarchy, identified earlier.

Of particular relevance also is Simon's description of organisational structure of decision-making:

> an organisation can be pictured as a three-layered cake. In the bottom layer, we have the basic work processes – in the case of a manufacturing organisation, the processes that procure raw materials, manufacture the physical product, warehouse it, and ship it. In the middle layer, we have the programmed decision-making processes, the processes that govern the day-to-day operation of the manufacturing and distributive system.[14] In the top layer, we have the non-programmed decision-making processes, the processes that are required to design and redesign the entire system to provide it with its basic goals and objectives, and to monitor its performance. (1965, p. 98)

However, formal structuring of managerial decision-making was typically absent in the early days of industrial firms, described by Chandler. The entrepreneurs concerned themselves with all sections of their business, and as a result found themselves spending the bulk of their time in functional activities. Chandler specifically identifies two activities (a) supervision of operations, which effectively involves management of second-level activities in Boulding's taxonomy; and (b) buying and selling – programmed decisions involving routine, repetitive operations. While Chandler does not discuss the mechanics of purchasing materials and distributing goods, such activity may be programmed using third-level mechanisms. Long-term planning and non-programmed decision-making (fourth-level activities) were not a regular consideration of entrepreneurs.

The U-form reorients the decisions undertaken by allocating decision-making by hierarchical levels. Routine second-level operations become the responsibility of supervisors in respective functions, while managers become responsible for interfunctional co-ordination as heads of functional divisions (Williamson, 1970, p. 19). Interfunctional co-ordination and also overall enterprise direction is the responsibility of central office (Williamson, 1970, pp. 18, 111). Managerial subsystems are thus created in an overall hierarchy with specialisation of responsibility in decision-making. While co-ordination of the internal environment of the firm may require occasional unprogrammed decision-making by a managerial subsystem, the internal operations of the firm are under the control of management to a large degree, and consequently decision-making in a mature firm by a managerial subsystem is typically a highly programmed third-level activity, such as is exemplified by management by exception, or variance analysis.[15] Programmed decision-making techniques such as standard economic analysis and operations research may also be utilised to deal with routine problems concerning internal allocation of resources.[16] However, strategy is generally the responsibility of central office, with functional heads operating as advisers or advocates.

Consequently, functional heads are mainly concerned with third-level mechanisms for control of internal environments, with some fourth-level activity, while central office concerns itself with strategy (fourth-level activity) and interfunctional co-ordination (third- and fourth-level activity). Complexity of decision-making increases progressively up the managerial hierarchy. The higher levels of management are typically concerned with higher levels of decision-making.

However, even this hierarchical arrangement was not sufficient. Expansion of the U-form, especially through diversification, created problems for central office; operating decisions began to demand more and more time of the scarce resources of central office. Creation of a further rung in the hierarchy through development of the M-form permitted the highest managerial subsystem to concentrate on the crucial fourth-level strategic problems while co-ordinating overall control of the enterprise, typically through lower-level decisions.[17]

Thus the development of the corporation has been directed towards increasing specialisation of the top managerial subsystem and higher-level decision-making. In the early days the entrepreneur was mostly concerned with second- and third-level operations and decisions, with occasional fourth-level decisions. The intended orientation of senior management in the U-form was third- and fourth-level decision, though increased demands from operating problems resulted in fourth-level decision being gradually 'squeezed out' of the system.[18] The senior management of the modern M-form are mostly concerned with fourth-level decision, with attention also being paid to third-level decision. This is of crucial importance since, as we have seen, the behavioural theory of the firm is primarily concerned with third-level model-building. As we would expect from the argument of Chapter 3, it is therefore incapable of modelling adequately the decision-making process of the top managerial subsystem. This problem will be the concern of the next chapter.

Before going on to look at the nature and organisation of R & D in the modern corporation, it may be a useful introduction to emphasise the process whereby the M-form organisation developed. A central feature of this development, the elaboration of hierarchical organisation, may be attributed to two complementary sources:

(a) *Natural differentiation.*[19] Differentiation is a property of open systems in which the open system evolves from a general and homogeneous state to one of increasing specialisation and hierarchical order. The development of programmed regulatory mechanism (third-level systems) is a feature of differentiation in biological and social systems.[20] The development of management hierarchy, from diffuse unstructured entrepreneurial activity in the nineteenth century through the U-form to the highly structured M-form, illustrates the process of differentiation.
(b) *Environmental change.* The development of turbulent fields and rapid environmental change has necessitated that top management should be free to consider strategic action. The M-form development was catalysed by the problems of dealing with environmental problems in the U-form.

However, differentiation involved qualitative changes in the way that decision-making was organised and oriented as well as facilitating certain types of decision-making. A feature of this process was an increasing tendency to treat technological change as endogenous phenomena capable of management and incorporation in the firm. The institutionalisation of technological change is now a feature of modern corporations, and it is this development in the context of the differentiation of corporate hierarchy which is of prime importance in contemporary analysis of corporate behaviour and change.

THE R & D FUNCTION AND THE INSTITUTIONALISATION OF TECHNOLOGICAL CHANGE

The pace of technological change in the nineteenth century tended to be dictated by individual inventors and perceptive entrepreneurs who appreciated the opportunities offered by a novel idea. Although in a number of cases there was co-operative development of invention by a small informal team, the individualistic role of the entrepreneur in that period tends to find a parallel in that of the inventor.[21] Invention tended to be exogenous to the firm, but with the development of the multifunctional firm, interest in developing innovations within the corporation increased, and scientists and technologists began to be employed in this capacity. However, in an analysis of the historical development of industrial R & D, Sanderson (1972, p. 139) points out that there was only limited development of 'in-house' R & D before 1914, the burden of discovery still resting on the individual inventor and such contact that he could establish with universities and within his own firm.

Sanderson identifies this as a transitional stage: 'The rise of the research department within the firm in the 1920s and 1930s not only changed this by bringing invention into the firm, divorced from the person of the entrepreneur; it also created a new situation whereby new developments came not so much from one firm or one research team as from the cumulative advance of knowledge through the interaction of many of them together' (p. 139). Thus, widespread institutionalisation of R & D coincided with the development of M-form organisational structures; from Sanderson's interpretation it also appears to have significantly contributed to the subsequent development of turbulent fields.[22]

However, in another historical analysis of the development of R & D laboratories, Fusfield (1975, p. 13) identifies four stages of development of industrial R & D after 1930. During the first phase in the 1930s the main focus of research management's attention was inside the laboratory, its main concern being with efficient management of research funds. The second phase developed in the 1950s, though some evidence suggests it was beginning to stir before then; this might be termed the integrative phase, in which the interaction of the R & D function with other parts of the organisation began to be emphasised. The third phase developed in the 1960s in which attention began to be focused on the impact of the environment on industrial research and vice versa. Fusfield also identifies a notional embryonic fourth stage in which not only the environment of industrial research becomes of importance, but also the question of the kind of society in which private enterprise can exist.

Therefore, R & D has become an institutionalised and integrated sector of corporate activity. However, it has also developed both the complexity of its relations with the rest of the firm and the external environment, and progressively extended the degree of openness of those relations. Although R & D become internalised in the corporation during the course of the twentieth century, initially there was minimal coupling between R & D and its environment.[23] The integration of R & D in the firm was increased through improved intra-firm coupling in the post-war period, while latterly there has been evidence of increasing coupling between R & D and its external environment[24].

Typically, R & D is organised on an auxiliary department basis within the M-form. It acts as a service department to central office and thus is directly responsible to the office in charge of strategic planning without having to go through divisional heads. Marris (1971, pp. 276–7) comments that while some R & D activity may be organised on a divisional basis, it is fundamental to the nature of R & D that it is likely to lead to the establishment of new operating divisions, and consequently tends to be organised on a separate, functional basis reporting to general office. Marris uses the term 'transcendent' to describe M-form corporations which perceive the possibility of changing their environment (by implication, open systems). 'Development' consists of conventional R & D and other strategies related to diversification or merger. Marris's model describes open system M-form behaviour in which R & D is a high-level activity.

Chandler in fact identifies the institutionalisation of diversification as being the consequence of the formation of a general office in the M-form and the development of an R & D department. He points out that science-based industries have increasingly oriented strategy towards developing products to ensure profitable use of resources that are increasingly technology-rather than product-based. The R & D department develops and analyses new products, while executives in general office are free to concentrate on whether the new product prospects and use of resources justify production. General office also decides whether production and sales can be handled through an existing division or whether a new division should be created (Chandler, 1962, p. 490).

In matters of strategy involving a great deal of resource commitment, R & D heads usually act in an advisory rather than an executive capacity (see Chandler, 1962, pp. 130–1, 354, 362–3, 376, 490).

Thus the development of R & D as a specialised, integrated function in the industrial corporation has been complementary to the widespread adoption of the highly structured and functionally differentiated M-form. The development of turbulent fields has been both a consequence of such developments and a catalyst for further evolution of the R & D function and M-form organisations. Research and development is a prime originator of strategic change in the firm; if the post-war development of turbulent fields in industry generally had created insuperable difficulties for the formulation of strategy by the top managerial decision-making subsystems, we would expect to find that growth of industrial R & D would have been inhibited and even reversed by the existence of such dynamic and uncertain environments. However, N.S.F. statistics indicate that this did not happen (see, for example, N.S.F., 1973a, p. 14, table 3). There was a decline in R & D staff employed in industry after 1969, but this is largely accounted for in terms of reduced federal support for defence and space programmes.

Therefore, while dynamic and uncertain environments are a feature of the context in which post-war strategic decision-making in corporations is typically set, and while development of R & D would itself be expected to contribute to the degree of turbulence of those fields, industrial corporations had tended to increase their involvement in this area rather than restrict R & D spending as a reaction to turbulence of environment. While turbulence may affect the formulation of R & D strategy, it is obvious from the growth of R & D spending that it has not prevented the corporation from increasingly relying on technological innovation as an active source of strategy options, and indeed we would expect that this reliance may increase through turbulence even though the complexity of the problem may also be magnified. Open system decision-making and planning is a feature of the modern firm, and in the context of turbulence its absence tends to create myopia and consequent vulnerability to unanticipated changes in environment, imperilling the survival of the corporation.

In the structurally differentiated firm, R & D therefore tends to become institutionalised as an auxiliary subsystem. However, institutionalisation itself changes the character of R & D activity; R & D as a function is a manifestly different concept from R & D projects. R & D function implies the continuing existence and employment of durable resources – plant, machinery, researchers, technicians, etc. On the other hand, at least in retrospect, an R & D project may be regarded as a series of operations or acts, the progression of which may be analysed by reference to the information generated at each phase or stage of the project. Resources, individually or in combination, contribute in varying degrees to the project development. However, while the R & D function in the corporation may be effectively distinguished from other functions and specified by reference to the resources employed, the defining characteristic of the R & D project is informational. The effect of resources on the R & D project is measured by contribution to derivation of information at each stage. As was emphasised in Chapter 2, such effects are by definition transitory and non-repetitive, creating unavoidable uncertainty and problems for rational decision-making.

Yet the institutionalisation of R & D constitutes a radical re-orientation of the focus of the M-form. In this latter respect, Bower was earlier quoted as describing the M-form innovation as creating a 'new order of complexity' in the firm. In a sense the opposite is true; essentially both the M-form and institutionalisation of R & D do not create complexity, but adapt to its existence. Both might be regarded as creating a new order *in* complexity, imposing structure and pattern on dynamic, disordered situations and processes, at least as far as the organisation and allocation of resources to those areas is concerned. The difficulties of operational goal specification, co-ordination of activities and separation of strategic and operating decisions in the expanded U-form enterprise[25] were largely alleviated by further hierarchical differentiation of the enterprise in the M-form. The institutionalisation of R & D may be simply regarded as a feature of the development of specialist activities on the part of the corporation in an attempt to deal with the complexities of its internal and external environment.

However, beyond simple specialisation, institutionalisation of R & D involves changes in the characteristics of the relevant decision-making variables. As mentioned above, the R & D project evolves through intermediate transient states of incomplete information, while the R & D function is specified in terms of stable resources of varying though generally substantial durability. The inimitability and consequent uncertainty of R & D projects discussed in Chapter 2 results in the well-documented difficulties of rational analysis. The R & D function, on the other hand, is specified in terms of resources; this raises the crucial question of how allocation of resources and decision-making is carried out in such circumstances. R & D imposes a new order on the organisation of technological change, and if this order is paralleled with identifiable regular and stable decision rules, it offers the promise and opportunity of a new direction for decision analysis and model-building for technological change. In the next section this area will be examined further, and in particular attention will be concentrated on conventions for setting R & D budgets in modern corporations.

MANAGERIAL PREFERENCE FOR R & D ACTIVITY IN THE FIRM

From the project-based neoclassical theory of the firm and its extensions, the problem of optimising involves the estimation of expected future streams of cost and revenue from projects and combinations of projects, and the calculation of that combination of projects which will maximise expected future profit. The R & D budget would not be calculated directly since it is not an explicit component of the optimisation problem, but would be calculable *ex post* as the sum total of selected projects identifiable as R & D projects.[26] The allocation of resources to R & D is consequently determined by the number of approved projects.

In fact this rational model of allocations, with its associated problems of unavoidable uncertainty, is rarely, if ever, used in large modern corporations. Allen (1970), in a study of R & D budgeting in U.K. firms, found: 'Usually authority for determining the R & D budget, and how the research should be financed, rested outside the R & D department, many organisations considering the annual research budget as a single item' (p. 176). Mueller (1966) concluded from a series of interviews held with research directors: 'Directors first seem to settle on a figure for the total and then divide it into its components' (p.37). Kaplan (1959) found that research directors frequently used constraint of available funds as a rationale for restricting or stopping projects (p. 33). Reeves (1958), who had responsibility in research budget formulation in the Esso Research and Engineering company, states: 'a long-range budget is not a project budget in which we outline year by year exactly what we shall be doing. It is, instead, a forecast of the over-all magnitude of effort that will be required and of the general type of work that will have to be done' (p. 136). Reeves states that in the development of a long-range research budget a basic assumption is that a lack of good ideas will present no problem even if those ideas are not available at the time of budgeting. Collier (1964), in an analysis of factors that define upper limits on research spending, found: 'The first of these is concerned with a balance of the total company resources' (p. 404), thus explicitly changing the focus from projects to R & D as a function.

These studies suggest a radically different picture of the R & D budgeting process than is suggested by the neoclassical theory of the firm. In the project-based optimisation models the R & D budget is simultaneously determined with project selection. However, the above studies imply the separation of budgetary and selection/control decisions; the budget generally precedes consideration of the direction of research. The separation of these two decision areas has consequences for the behaviour and attitudes held by R & D administrators and researchers. Kaplan (1959) points out that research administrators can, and do, tend to blame funding and procedural problems on senior administrators external to the functions, while still themselves retaining control of the substantive issues of technical direction. Allen (1970) found that although authority for deciding the R & D budget lay outside R & D departments in most cases, in almost all the firms studied allocation of R & D resources was at the discretion of the R & D leader or director.[27] Jewkes et al. (1969) found from their study of invention that while it was possible to organise and plan for the provision of facilities and resources to R & D, co-ordinating teamwork and guiding research activity involved problems of a substantially greater order of magnitude (pp. 107–14).

The R & D budget is therefore generally based on consideration of the resources which should be allocated to R & D rather than specific projects. The top managerial decision-makers are free to consider in the light of past experience how much should be allocated to R & D relative to expenditures on other functions. Freeman (1974) points out: 'Although management cannot calculate accurately the return on any individual project or piece of R & D, it has learnt from experience and from observation of competitors that this "normal" level of R and D spending will probably help it to survive and grow' (p. 246). This suggests that resource allocations to R & D are potentially more stable than might be expected if consideration were limited to study and analysis at project level. Gold (1971), from an extensive review and synthesis of case studies and analyses in this area, provides support for this in his 'basic hypothesis' that 'in most firms top management tends to have a reasonably stable preference for the means of promoting its primary objectives (such as improving or maintaining: profitability; growth; market position; security of assets; relative stability of operating levels; and a favourable public image)' (p. 222). Technological progress is not seen as a primary objective in itself, but one of the possible means of promoting more fundamental goals.[28]

Empirical analysis of allocations to R & D at industry level by

Freeman (1962) indicates that R & D expenditures were stable over a number of years, while Mueller (1966, p. 36) refers to the evidence cited in his unpublished Ph. D. dissertation (1968) which indicated that 'R & D data, unlike patents, are not subject to erratic year-to-year fluctuation'. In fact, stability of preference for R & D resources, together with the concern expressed by Collier in achieving a balance of total company resources, may help to explain what Schmookler (1962b, p. 213) identified as 'the common corporate practice of setting research and development budgets at a fixed percentage of sales'. This is a widely reported practice used in determination of the research budget.[29] It has also been reported as a practice in allocating funds to marketing, a function which, like R & D, typically involves projects with associated uncertainty as to future returns.[30]

This is interpretable in terms of a further defining characteristic of open systems, that of the steady state. According to Katz and Kahn (1966): 'open systems which survive are characterised by a steady state. A steady state is not motionless or a true equilibrium. There is a continuous inflow of energy from the external environment and a continuous export of the products of the system, but the character of the system, the ratio of the energy exchanges and the relations between parts, remains the same' (p. 23).[31] It would seem that steady state is identifiable at least in two respects for many corporations, firstly at a cognitive level in terms of managerial preferences, and secondly at an allocative level in terms of observed deployment of resources to R & D.

A further element contributing to steady-state behaviour in the R & D function is perceived need for stability of employment of R & D resources. In a study of 43 chemical companies in six European countries, Olin (1973) found: 'The most important factor in determining the size of the R & D budget is the last year's expenditure. It was generally thought to be detrimental to the R & D effort if it was changed by more than 5–10% annually in any direction' (p. 127). Mansfield (1968a, p. 62) states that firms place emphasis on the stability of their R & D programme; they attempt to build up gradually and avoid expansion which may soon have to be cut back.[32] Reeves (1958) justifies Esso Research and Engineering's research budgeting based on 'magnitude of effort' rather than project analysis, since 'a research organisation is an extremely complex machine and takes far longer to build than most other corporate assets such as a new factory' (p. 136). Reeves also states that high variability in research budgets can be highly disruptive in organisational terms.

Thus, stability of preference for R & D resources, and stability as a desirable feature of the R & D function facilitating continuity and constancy of employment of R & D resources, may be complementary to one another and mutually reinforcing. Cyert and March (1963) also emphasised the inherent stability of the R & D function on the basis of research conducted by Seeber. Cyert and March conclude that there are four important features of the R & D process:

(i) Most organisations are aware of and probably use such simple rules as per cent of revenue as a guide to research and development allocation.
(ii) The pressure on subunits for maintaining absolute dollar allocations, the logic of research appropriations, and the difficulties of forecasting revenues lead to considerable attempts to smooth allocations so that they vary less from year to year than do revenues.
(iii) Target allocations are substantially influenced by estimates of allocations (per cent of sales) in other "comparable" organisations.
(iv) Organisational failure on profit or sales goals leads to pressure to revise the allocation rules.' (pp. 274–5).

Points (i) and (ii) emphasise simple and intendedly stabilising resource-based decision-making while points (iii) and (iv) provide evidence of the environmental phase earlier identified by Fusfield (1975). Point (iv) indicates the role of negative feedback in adjusting steady-state preferences. Particularly important is the tendency found for stability in resource allocation in the face of variability in the revenue from projects. This is explained by Freeman (1974): 'many R and D managers or scientists act as if they were farmers. They know that there are unpredictable and accidental factors present in their work. But they also know that, if they apply their labour with ingenuity and appropriate equipment over a sufficiently long period, they will probably achieve some useful results' (p. 339). Freeman suggests that the existence of commercial research institutes and the steady increase of company-financed R & D are evidence of the economic practicability of a wide range of types of R & D activity, even if particular projects are inherently uncertain.

Therefore, stability of resource allocations does not necessarily imply regularity and predictability of R & D output. While this may appear obvious, it is frequently ignored or confused; thus, Chandler (1962) states: 'the systematising of strategic decisions through the building of a general office and the routinising of product development by the formation of a research department have, in a sense, institutionalised this strategy of diversification' (p. 490).[33] Yet a viable R & D department with stable and continuous use and employment of resources need not produce 'routinised innovations' – indeed, it was argued in Chapter 2 that the concept is implicitly contradictory. The tendency for management to exhibit stable preferences for allocations to R & D does not necessarily imply expectation that there will be a corresponding steady yield of commercially useful ideas. The R & D budget decision is typically separate from allocative and control decisions in the R & D function, and concerned with radically different decision variables. Stability in budgeting is entirely consistent with highly uncertain and unpredictable R & D projects.[34]

Bower (1970), in analysing conventional resource allocation decision-making in modern firms, reinforces the idea of qualitative differences in decision variables operative at different levels. Typically, top management is concerned with the overall relationship of the firm with its environment while subunits are concerned with the specific content of overall plans (p. 190). This is consistent with the argument of this chapter, and Bower emphasises that 'the process shaping the content of plans – both the choice of objectives and the discrete commitment of resources – is different from the process that leads the plans to be approved' (p. 190). The subunit manager 'plans within the scope of his job as it has been defined' (p. 190).

The significance of these analyses may be regarded as twofold. First of all, a number of contentious problem areas in the study of R & D, still subject to active debate and involving difficulties of analysis, may be subsumed into a separate area of analysis. The high degree of uncertainty associated with specific R & D projects and difficulties in formulating long-term strategy in turbulent environments are specifically problems of internal control and management. While they may have implications for R & D budgeting, the decision problem at this level is qualitatively different from the management and control problem, and consequently the relevance of issues must be reappraised in the context of this contrasting perspective. The R & D system as a whole may display properties which are not apparent from consideration of its component parts or processes.

Secondly, and related to the above, stability and institutionalisation of R & D resources emphasises the integration of technological change as a means by which the firm can deal with its environment. It emphasises the allocation of resources for the contrivance of strategies and so stresses the need to regard the firm as an open system. The differentiation in functional hierarchy associated with the evolution of the modern corporation, and the tendency for allocation of resources to the R & D function to exhibit 'steady state' behaviour relative to allocations to other functions, are both characteristics of open systems behaviour. It provides support for the earlier suggestion that the behavioural theory of the firm may need revision or augmentation to provide for fourth-level activity in the firm.

It must be emphasised once more that these arguments relate to the large modern corporation, particularly those operating in a regime of rapid technological change. Studies reporting contrary findings to the consensus do exist, and these will be discussed in Chapter 6. The dissenting studies will prove useful to the present analysis, since as exceptions they effectively test the rule of stable resource preferences.

In addition to the possibility of exceptions to stability of preferences at functional level, it is conceivable that at lower levels than that of the function there may be activities having in common characteristics of a systemic nature. In such circumstances, as with the overall system, it may be possible to identify stable and organised properties at a higher level than that of individual projects. In this respect, Chapter 2 indicated the possibility of identifying research and development subsystems arranged in a hierarchical interlinking ordering with respect to final output, and provides a basis for further analysis of this possibility.

If stable preferences for R & D subsystems are identifiable, it may indicate how model-building of corporate behaviour may extend the application of the concept of a resource preference system to levels below that of the overall function. In particular, it would suggest the possibility of a hierarchy of 'top-down' resource allocation, in which resource allocations at lower levels compete with other uses of funds within the overall constraint of a function budget. The function budget, while operating as a variable at higher levels in the decision-making process, would operate as a parameter at these lower levels. The lower-level allocations within the overall budget would then in turn provide constraints within which further sub-allocation is made.

However, even if resource allocation were made in this fashion, there must of course be a level of allocations sufficiently low in the decision-making hierarchy where allocations are made in terms of projects, not resources. In the next section it will be suggested that there are qualitative differences in managerial attitudes towards, and preference for, different types of R & D subsystem activity.

MANAGERIAL PREFERENCES AND R & D SUBSYSTEMS

In Chapter 2 distinction was made between types of R & D subsystem: basic/fundamental research, applied research/inventive work, and development. The linkage between subsystems, and their hierarchical relationship to final output, contributed to the complexity and uncertainty of evaluating expected final output from inputs into a particular subsystem. We shall consider both sets of interpretation of R & D subsystems as delineating a 'spectrum' of R & D subsystems, with basic/fundamental research at the 'top' end, and development at the 'bottom' end.

As Freeman (see Table 2.1) indicates, the further that the level in which an R & D project is operating is removed from final output, the higher the degree of uncertainty associated with projects operating at that stage. The reasons for this were discussed in Chapter 2. However, there is typically a further distinguishing feature of final output derived from respective subsystems: the degree of radicalness associated with innovation originating from different levels in the subsystem hierarchy tends to increase as activity moves towards the 'top' end of the spectrum. It may be that the output of one applied research project may stimulate successive development projects designed to differentiate moderately or improve the derived innovation or innovations, and consequently R & D is active only in the development subsystem; in those circumstances the final output in terms of products and processes will tend to be only moderately or marginally different from existing output, as indicated by categories 5, 6 and possibly 4 in Freeman's table. Similarly, it may be that applied research may be conducted from a fixed scientific base using a given stock of scientific knowledge; however, in this case new technology and inventions will tend to be highly novel and radical with regard to existing output. Inventions may require major revision of productive techniques and may require further inventive work before viable innovations are obtained, creating complex technological organisational problems of integration in the existing product/market posture of the firm. This is typified by categories 2, 3 and possibly 4.

However, potentially the most radical and fundamentally different innovation involves a shifting stock of scientific knowledge and implies activity in the basic research subsystem as defined earlier by Machlup. Nelson (1959a, p. 302) claims that significant innovations have tended to come from basic research, not applied research, and gives examples of X-ray analysis, radio communication and the development of hybrid corn, each of which was facilitated or made by scientific research. Freeman (1974, p. 261) identifies in-house fundamental research as being contributory to the development of nylon, polyethylene and the transistor among other inventions with far-reaching significance and impact.

As would be expected from the discussion in Chapter 2, the degree of radicalness of final output, measured in terms of qualitative differences compared to existing output, tends to parallel the degree of uncertainty associated with the potential utility of inputs in the particular subsystems. Thus, contribution to degree of radicalness of innovation also tends to increase as R & D activity moves towards the basic research end of the hierarchy. The nearer the 'top' end of the spectrum, the greater the degree of uncertainty associated with derived final output, and the greater the degree of radicalness typically associated with final output (see Fig. 4.2).

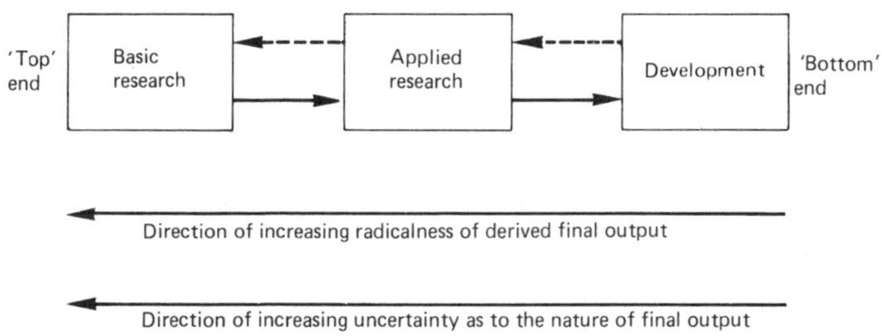

FIG. 4.2.

As we shall discuss further below, management is not indifferent to these two influences. Further, since both appear to achieve greater significance the nearer the relevant activity is to the 'top' end of the R & D spectrum, we might expect management to rank their preferences for particular types of R & D activity according to their position in the R & D spectrum.

In order to establish such ranking, we need to examine the effect of uncertainty and radicalness of final output on managerial attitudes. While such attitudes are linked and frequently mutually reinforcing, it is possible to separate out distinct influences on managerial perception of R & D subsystem characteristics, and consequent attitudes to allocation of resources to those areas.

Dominant influences include the following:

(i) *The pervasiveness of the rational model and its incompatibility with uncertainty* (see Gold 1971; Schon, 1967). The rational model is appropriate as a general frame of reference for stable environments and routine decision-making problems. However, the greater the degree of uncertainty associated with a project, the more likely that managers adhering to the rational model will screen such projects out of the decision system. Uncertainty in R & D is derived from two project-specific sources:

  (a) Technical uncertainty; this tends to increase as R & D activity moves to the 'top end' of the R & D hierarchy (see chapter 2).

  (b) Market uncertainty; this is a consequence of the development of turbulent fields in the industrial environment.[35] The uncertainty is compounded back by the length of time from project inception to innovation in turbulent fields, and also by the degree of novelty and unfamiliarity of innovation, and in both respects uncertainty again increases towards the basic research level.

(ii) *Resistance to radical innovation.* This is generally contingent on two related sources:

  (a) The difficulties of co-ordinating and integrating radical innovation in complex systems: 'The problem with trying to achieve major advances in large and complex systems-products with a large number of highly interdependent components is that to change any one item causes reverberations throughout the system' (Nelson et al., 1967, p. 27).[36] Administrative problems in implementing innovation in richly connected, complex and functionally differentiated systems tend to increase with the radicalness of the innovation, the disruptive potential of radical change often acting as a strong dissuasive influence.

  (b) Dynamic conservatism (Schon, 1971, pp. 30–57). This is active resistance to external or internal threats to stability on the part of individuals and groups in the system. In this case it is not the complexity of radical change which is the source of difficulty, but the extent of the threat to the stability of the existing system, the *status quo*, and vested interest. In his earlier work (1967) Schon identified the rational model as a crutch and reinforcing agent of dynamic conservatism.

Again, since degree of radicalness increases towards the 'top' end of the spectrum, we would expect discrimination against, and resistance to, R & D subsystem activity to increase with the level of the subsystem in the R & D hierarchy.

A further factor encouraging resistance to a particular type of subsystem activity is that in the case of the basic research subsystem we would expect non-appropriability of subsystem output, referred to earlier, to reduce its ranking in a managerial preference system even further.

(a) and (b) would tend to encourage increasing managerial discrimination and bias against R & D activity as the level of the relevant R & D subsystem approaches the basic research end, while the third influence reinforces the tendency for basic research to be a lower-ranking subsystem in managerial preferences.

All these effects would tend to influence management in the same direction, biasing them against innovation activity towards the 'top' end of the R & D spectrum. This is supported by Gold (1971), in the light of his review and analysis of studies of innovation activity, in that as far as top management is concerned:

> first preference is generally for the continuation, or only moderate intensification, of familiar operations involving little risk to established organisational structures or patterns of resource allocations . . . the generating or pioneering adoption of major technological innovations is likely to rank low because it tends to involve heavy investments, substantial risks and readjustments in existing organisational arrangements and budgeting allocations affecting many functions and operating divisions. (p. 222)

Jewkes et al. (1969, pp. 112–14) concluded from their case studies of invention that administrators tend to discriminate against uncertain and long-term research projects; research departments themselves may be a source of resistance to change due to the N.I.H. ('not invented here') syndrome.[37] Hamberg (1963) described most corporate research as being mainly concerned with unambitious and 'safe' projects, while Nelson et al. (1967) found in interviews with corporation executives that R & D activity tended to be concentrated in search for modest improvements and short pay-back periods.[38] These observations support the view that management tend to prefer less uncertain and less novel projects as opposed to highly uncertain and radical projects.

Since both the radicalness and uncertainty of derivative final output tends to increase from the 'bottom' to the 'top' end of the R & D spectrum, this supports a general hypothesis: *management will rank their preferences for a particular type of R & D activity according to its position in the R & D spectrum; the closer a particular activity is to final output, the less liable are management to resist or discriminate against that particular activity.*

This suggests that there may be obstacles to be overcome before activity towards the 'top' end of the R & D spectrum is taken up. In particular it raises the interesting possibility that there may be progressively strengthening resistance to specific R & D activity as we move from the 'bottom' to the 'top' of the R & D spectrum, represented by ever-increasing barriers or thresholds which must be overcome before the activity can be conducted. In Chapter 7, in a modified version of the framework to be developed in the next chapter, we shall consider those factors that may influence whether or not a particular barrier to the uptake of a lower-ranked activity may be overcome.

One relevant question which has not been discussed so far is whether or not allocation of resources to R & D subsystems has inherent stability comparable to that of the R & D system overall. This is not an issue which appears to have received much attention, but a recent report by the National Science Foundation (1973c) provides relevant evidence at the level of the firm based primarily on surveys of senior R & D officials in 55 of the largest U.S.corporations. The officials were asked to comment on possible effects of cut-backs on basic research; the authors concluded:

> It was generally believed .... that such a policy on a long-term basis could spell disaster yet, in periods of economic duress, cutting back on basic research can be an effective short-term mechanism to improve profits. If the basic research cut-back is too deep, however, good research people must be released; it is extremely difficult to obtain suitable replacements when money again becomes available. (pp. 3–4)

In the previous section, Olin (1973), Mansfield (1968a) and Reeves (1958) emphasised the general belief held by corporate management that high variability in the overall R & D programme can have an adverse and disruptive effect. The N.S.F. study above suggests that this attitude may still hold at subsystem level, in this case basic research. While the vulnerability of basic research to short-term pressure is emphasised, so also is the resource-based perspective of managers and inherent tendencies towards stability of basic research allocations. We would expect potential vulnerability of basic research in distress conditions to be high owing to the irreconcilability of basic research and the rational model, and the long-term nature of basic research payoff. However, it would seem that in the absence of short-term pressure basic research has a tendency towards a stable share of total resource allocation within the R & D department.

SUMMARY

In this chapter we have looked at some trends in the evolution of the modern corporation. The process of differentiation, reflected here in the development of the M-form organisation, has played a central role in the changing nature of the corporation in the twentieth century. As the environment of the typical modern corporation has become more complex and uncertain, so the increasing need for effective strategy has been paralleled by increasing difficulties and obstacles to such formulation.

However, it was suggested that the apparently widespread existence of stable managerial preferences for corporate resources in the various subsystems meant that analysis and selection of individual strategies and projects could be disregarded as far as the present analysis is concerned. The institutionalisation of R & D and the separation of the overall R & D budget decision from project selection in the large modern corporation indicates the manner in which such preferences have evolved. The possibility that stable managerial resource preferences may exist at intra-functional levels was also considered, the specification of possible subsystems (such as 'basic research') being suggested by the analysis of Chapter 2.

A basis for an alternative to the highly unsatisfactory project-based model-building of approaches such as the neoclassical theory of the firm is indicated by the evidence that management of large corporations commonly seek to achieve a balance of resources in corporate operations. Cyert and March (1963) recognised that it may be useful to discriminate between activities or projects and subunits or subsystems in the firm and, as we have seen in this chapter, they provided simple behavioural principles relevant to the analysis of subsystems rather than component activities. However, as we have seen earlier, their analysis covers a restricted class of decision problems and cannot be satisfactorily applied to top managerial decision-making in the M-form organisation. We shall therefore use the general conclusions of this chapter as the starting-point for the development of the systems framework of the next chapter.

5 The Framework

We are now in a position where we may begin to draw together the different threads of the arguments of preceding chapters with the intention of formulating a model of resource allocation in the modern corporation. So far we have been concerned with three interrelated aspects of the R & D problem, that is, the nature and characteristics of R & D, models of allocation of resources for technological change in industry, and the evolution of the modern corporation and institutionalisation of R & D. In this chapter the development of a systems framework for analysing corporate allocations to R & D is founded on a number of points and conclusions made in these earlier chapters.

A central issue dealt with in the previous analysis is the significance of uncertainty in R & D work at project level in the modern corporation. Because of the pervasiveness and importance of uncertainty, a substantial body of opinion, supported by empirical studies of R & D, holds that the applicability of rational models to R & D projects is drastically constrained. The tendency for large modern corporations to separate budget and allocation decisions reinforces criticism of project models; if, as appears to be the general case, corporations generally allocate a stable percentage of available funds and resources to R & D *independently* of the allocation decision, the irrelevance of project-based models is assured, at least as far as descriptive theories of the R & D budget in the firm are concerned.

If we retain a project-based perspective of allocation of resources to functions in the firm, such budgetary conventions are difficult to explain or justify. In fact, such a frame of reference encourages criticism in terms of the imputed non-rational and unsound nature of such a budgetary process. Quinn (1959, p. 295) describes the practice of setting the R & D budget as a percentage of sales as 'arbitrary', Jastram (1950) identifies the practice of setting advertising appropriation as a percentage of established sales revenue for the period, 'regardless of any logical inconsistency which may be involved' (p. 155), while Doyle also criticises such appropriation policy as lacking a logical basis (1968, p. 577). It is indeed difficult to conceive how such practice might be analysed or condoned within a project-based frame of reference.

Yet 'logical inconsistency' or 'arbitrariness' only holds as valid criticism if it is intended that budgeting adopt the frame of reference held by the same critics. It may be that such conventions follow a framework of internally consistent rules unrelated to project-based optimisation procedures, and consequently acting in accordance with an entirely different rationale. In fact in the last chapter it was suggested that the R & D budget decision is frequently based on consideration of qualitatively different characteristics and variables from those typically associated with project-based models. The widespread resource orientation of budget decisions parallels and complements functional differentiation in hierarchy and the institutionalisation of technological change. These developments are related to the growth of turbulent fields and the corresponding necessity for open system organisation and action on the part of the corporation.

The last chapter also described features of corporate organisation and behaviour constituting definitive characteristics of open systems, and further supports adoption of the open system frame of reference in analysis of technological change. The typical variables relevant to decision-making at higher function levels were identified as resource rather than the outcome-orientated project. Beyond mentioning differences in properties of those decision variables (particularly aspects of stability and homogeneity associated with resources and projects respectively), we have not examined the implications of this feature of higher decision-making procedure. In this chapter we attempt to provide a rationale for such behaviour, in an open system framework. We shall develop a framework for corporate resource allocation based on the concept of the firm as a hierarchically organised system, with a relatively stable set of preferences for resource allocation, distributing resources in a 'top-down' fashion. The firm is regarded as a system partitioned into subsystems; allocations are first made to individual systems and then distributed to component subsystems. A means whereby a preference system can be built up through feedback is also suggested. The corporate management will be assumed to allocate resources, according to this preference system, with the primary objective of ensuring the survival of the corporate system, in the context of a hostile turbulent environment (actual or potential).

The view of the firm as being hierarchically structured into successive partitionings was emphasised in the last chapter as a central feature of corporate resource allocation. In the next section we shall suggest how redundancy may facilitate, and non-reducibility necessitate, the description of the firm in this way.

ABSTRACTION IN INTERPRETATION OF COMPLEX SYSTEMS

In Chapter 1 it was suggested that a holistic view of the corporation may be more appropriate than the reductionist perspective of neoclassical theory for some purposes. Redundancy and non-reducibility were two concepts used to support the holistic or systemic argument, and they are discussed in rather more detail in this section. It is important to bear in mind that neither concept is definable in terms of the concepts and behaviour of lower levels. The previous chapter has, in fact, already suggested the emergence of stable resource preferences at higher levels in corporate decision-making, and implied that these are not reducible to project terms. We shall suggest why this may be the case below in discussion of 'Gestalt' organisation, but first the concept of redundancy is discussed as an additional justification for abstraction.

In this respect, H. A. Simon (1969) emphasises the possibility of deriving simple descriptions of complex systems sufficient to provide adequate models of behaviour: 'The more we are willing to abstract from the detail of a set of phenomena, the easier it becomes to simulate the phenomena. Moreover we do not have to know or guess at all the internal structure of the system, but only that part of it that is crucial to the abstraction' (p. 16).

Simon later asserts that complexity or simplicity of structure depends critically upon the manner of description. Most complex systems incorporate large degrees of redundancy, and this redundancy can be utilised to simplify description. However, the correct representation and simplification must be identified to achieve this (1969, p. 117).

An example is provided of the behaviour of an ant: 'An ant viewed as a behaving system, is quite simple. The apparent complexity of its behaviour over time is largely a reflection of the complexity of the environment in which it finds itself' (1969, p. 24). Simon develops this hypothesis from description of an ant's irregular, erratic path weaving and detouring around obstacles on the way to its anthole. Complexity of behaviour is determined by complexity of environment, not innate complexity of the ant itself in terms of goals or preferences.[1]

In this context the ant may be represented as a simple system: 'at the level of cells or molecules, ants are demonstrably complex; but these microscopic details of the inner environment may be largely irrelevant to the ants' behaviour in relation to the outer environment' (p. 25). Simon suggests that an automaton, though different macroscopically, might adequately simulate the ant's behaviour.

While Simon is careful to frame his interpretation as a hypothesis, his identification of redundancy of lower levels in the behaviour system parallels the gross simplicity of analytical and simulation models of complex systems.[2] Simon extends his hypothesis to cover human behaviour, and if it is further extended to social systems, it has obvious implications for corporate analysis. Irrelevance of study of behaviour at lower levels in the system would greatly simplify analysis. A corollary is that analysis of higher levels in terms of lower-level concepts may be inapposite and, worse, possibly misrepresent higher-level activity.

Consequently, characteristics and properties of systems may differ qualitatively from constituent subsystems. For example, steady-state behaviour may involve different perspectives or qualities from those of lower subsystems. This is pointed out in one respect by Rapoport (1974): 'Open systems if "left alone" tend toward steady states determined by their structure and interactions with the environment. Mathematically this trend is often demonstrated as a deterministic one. But it can also be shown in so-called *stochastic* systems in which the simple events are probabilistic but the system behaves deterministically in its totality' (pp. 167–8).

However, systemic properties may differ substantially from simple aggregative or summative properties derived from mathematical manipulation of statistical events. The Gestalt psychologists based their analysis on the hypothesis that wholes (which Angyal, 1969, describes as 'organised objects') have properties not derivable from analysis and aggregation of constituent parts. This approach has been widely summarised in terms of what Angyal (1969, p. 26) describes as an inapt formulation: 'the whole is more than the sum of its parts'. As Angyal points out, this may suggest that a summation of parts takes place and that in addition a new factor enters the composition of wholes, contrary to the interpretation of Gestalt psychologists. Gestalts and additive aggregations bear no direct relationship to one another; instead of implying that wholes involve something more than summation of parts, it is more appropriate to state that aggregation plays no part in whole formation (Angyal, 1969, p. 26).

This is explicitly formulated by Angyal: 'in an aggregation the parts are added, in wholes the parts are arranged in a system. The system is an independent framework in which the parts are placed. . . . the whole is, to a large extent, independent of the individual parts' (p. 27). Consequently, aggregation and the formulation of wholes involve processes of entirely different natures. 'Gestalt' is German for 'configuration' or 'pattern' and, as the name implies, holistic properties of systems may only be perceivable through consideration of higher levels in the system. Thus, not only would analysis of Simon's ant at the level of cells and molecules be redundant, it might obscure simple goal-seeking behaviour at higher levels. The abstraction involved in considering progressively higher levels in systems not only involves information loss in terms of description of lower-level subsystems, but may also result in a gain due to qualitative differences in the determinants and constituents of higher-level behaviour. As Simon elsewhere states, in dealing with complex problems involving persistent, systematic properties and characteristics, 'Man . . . is not only a learning animal, he is a pattern finding and concept forming animal' (1959, p. 272).

The importance of this for the subsequent model development in this chapter is difficult to overstate. The general direction of contemporary economics is with respect to derivation of aggregative relationships between subsystems and system: costs of production are summed to obtain total costs of production, aggregate demand and supply curves are derived from the behaviour of individual units. The concepts of cost, revenue, demand and supply remain essentially the same in whatever level in the overall economic system they are applied.

However, in allocation of resources to R & D and other functions in the large modern corporation, there does not appear to be a summative process and corresponding concept common to all subsystem levels. The project is naturally the operational concern inside the R & D system itself, but it has typically no equivalent in budgeting at higher levels in the large corporation. Budgeting at the level of the function is typically analysed in terms of available resources and precedes consideration of allocations to specific resources. Managers evince stable preferences for resources, and firms are generally describable in terms of enduring departments, managerial teams, etc. The corporate system as a whole, analysed in terms of consistent functions such as R & D, marketing, production and so on, does appear to be largely definable independently of its component projects and activities; not only is this a defining quality of 'resources',[3] it is also a feature of Gestalt organisation of wholes (see Angyal above).

Gestalt organisation of managerial preferences for resources at the highest level in the firm may help to explain the processes and behaviour discussed in Chapter 4. The abstraction of top management with respect to routine problems and complex interrelated processes at lower levels may facilitate formulation of Gestalts with respect to resources. It is a characteristic of Gestalt that configuration or pattern at higher levels is not derivable in terms of 'building up' from lower-level analysis; consequently, project analysis may be inappropriate at this higher level, as far as description of managerial overview of the allocative problem in the corporation is concerned.

The idea of 'Gestalt' is almost certainly an essential ingredient in the formulation of stable 'meta-rules' which Nelson (1972, p. 42) postulates as existing in turbulent environments. The specification of constructs to which these meta-rules might refer or be applicable possibly ranks among the major, though generally unrecognised, contribution of both Cyert and March, and Penrose. Thus, Penrose contends that resources may be defined independently of their use (1959, p. 25), while Cyert and March perceive the 'relatively independent calculation' of subunit budgets with respect to projects (1963, p. 272), and distinguish between subunit and project allocations (p. 274). In neoclassical theory such distinctions would be arbitrary with no behavioural implications; boundaries and characteristics of subunits would be defined in terms of aggregative constituent projects, and resources and subunits would only be recognised in terms of their derived services.

The distinctiveness of Gestalts from simple aggregations is emphasised by Allport (1955): 'Since the percept has an indissoluble unity it could not have been acquired in the first instance by adding sensory units "piecemeal". It might take a little time before the Gestalt appears; but when it comes its advent is sudden. It is all or none. Learning would therefore have to be a discontinuous rather than a continuous process' (pp. 124–5).

Analogously, in the case of the corporation, patterns inherent in the organisation and its environment are not derivable from simple manipulation or analysis of projects. Such processes may require higher levels of abstraction and synthesis by corporation management. Obviously such perceptual constructs cannot be directly observed, and must be inferred from the usefulness of models based on such concepts; however, evidence of search for Gestalt organisation in a highly innovative corporation is provided by Morton (1967) from his descriptive comments of Bell Telephone Laboratories in which he was Vice-President for Electronic Components:

> Wherever you run up against a complex problem – and in fact, I find it difficult to name problems in the technological world that are not complex – you look for patterns. In other words you look at it as a system. What do we mean by a system? First it is an integrated whole. It is not a bunch of disconnected parts. It is a structure of specialised

> parts, each of which has its own special function in the whole, and each of which is so coupled to the others that they act together toward a common purpose. We also maintain that the combined effectiveness of the persons making up this system is more than the simple sum of its parts. (p. 22)

Morton analyses A.T. & T., the parent company of the Bell Telephone Laboratories (B.T.L.), into 'only four basic functions' (p. 25), of which B.T.L. serves as the R & D department. This functional separation is perceived in a complex system of about 800,000 employees in 24 companies (pp. 24–5).

Such a view of the individual corporation appears reasonable, and indeed essential in the light of the discussion here and in Chapter 4. We shall develop a framework describing corporate resource allocation based on a similar interpretation of the corporation as that described by Morton above. In the next section the features of managerial decision processes concerned with the overall deployment of corporate resources will be analysed further in terms of Gestalt organisation, and the basic assumptions of the framework will be set out.

NATURE AND OBJECTIVES OF CORPORATE RESOURCE ALLOCATION

The analysis so far suggests that certain general statements can be made regarding corporate resource allocation at high levels of decision-making in the large modern corporation. We shall briefly summarise the most significant as far as our argument is concerned.

(1) It is based on *subjective* rather than objective decision rules. There do not appear to be widely recognised principles or methods corresponding to the decision-making process which would provide the same allocative solutions by analytical methods. The evidence of the earlier chapters suggests that subjective evaluation of allocations takes place through rule-of-thumb methods, rather than aggregation of individual projects by means of optimisation techniques.

(2) There appear to be *stable managerial preferences for resources* at top management levels in the firm. This is reflected in the tendency towards explicit equiproportional allocations at functional level over time despite the existence of turbulent environments.

(3) The modern corporation behaves as an *open system* in allocating and organising resources for the generation of strategies and innovations.

Research and development at function level is an institutionalised and integrated part of the managerial preference system.

(4) Managerial decision-making based on the preference system for resources is *rational*. This is the most speculative of the statements made here, and indeed the burden of argument referred to earlier emphasises the 'arbitrary', 'illogical' nature of conventions such as setting R & D as a percentage of sales. For the reasons stated earlier, such criticisms may be provisionally discounted since they are made from within the neoclassical frame of reference. However, it was suggested earlier that firms may learn the appropriate level of R & D that permits them to survive and expand their operations; this infers subjectively rational action on the part of senior corporate management even if analytic methods are unreliable at project level. The development of models of corporation behaviour would be made exceedingly complex without some form of rationality postulate as guideline.

We now have a number of building-blocks on which to develop a model of corporation decision-making behaviour. One area of neglect so far is the possible objective function operative at this higher level of abstraction. In neoclassical theory the objective function is profit maximisation, and recent extensions have included sales maximisation subject to a profit constraint (Baumol, 1959) and maximisation of managerial utility (Williamson, 1964). In a recent qualification of his early theory of managerial discretion, however, Williamson (1971) has suggested that the M-form corporation may effectively promote profit-maximising behaviour.[4] We shall discuss this further below.

An alternative motivating force applicable to resource management at the top level in the managerial hierarchy is required. This is suggested by Emery (1969) in his analysis of corporations as open systems operating in fields with 'complex environmental interactions' (p. 8):

> The task of management is governed by the need to match constantly the actual and potential capacities of the enterprise to the actual and potential requirements of the environment. Only in this way can a mission be defined that may enable an enterprise to achieve a steady state. However, the actions of management cannot in themselves constitute a logically sufficient condition for achievement of a steady state. (p. 10)

Here the steady state is the target for managerial allocations. A similar objective is implied in Chamberlain (1968):

> There must always be a tendency toward a state of equilibrium. At the same time there must also be a tendency toward a break-up of existing relationships and the formation of new ones because of the intrusion of unavoidable environmental changes and the firm's purposiveness with respect to them.
>
> These two tendencies – toward coherence and disturbance, toward equilibrium and disequilibrium – must run together, in a kind of economic counterpoint. (p. 10)[5]

Chamberlain suggests that a balance or harmony must be maintained between existing relationships and forces promoting change, in which latter category would be included R & D. This is similar to Emery's primary task of management, in which achievement and maintenance of a steady state is the objective. The steady state was earlier interpreted as maintenance of the essential character or identify of the system; Kast and Rosenzweig (1974) reproduce this interpretation, but also provide a description of steady state appropriate in this context: 'The steady state has an additional meaning; within the organisational system, the various subsystems have achieved a balance of relationships and forces which allows the total system to perform effectively' (p. 116). A more rigorous definition is provided by Von Bertalanffy (1973, p. 167) as *a time-independent state in which the ratio of the system components is constant.* We shall refer to this description in the development of the framework.

This balance of subsystems and consequent 'matching' or 'harmony' between firm and environment may therefore be regarded as a form of organisational objective. Thus, organisational success in achieving this objective may be regarded as being reflected in the coincidence of the selected allocations of the managerial resource preference system with the appropriate allocation of resources required to achieve/maintain steady state. Such coincidence would result in observable equiproportional subsystem allocation of resources over a period of time. However, a mismatch between preference system and required subsystem balance would eventually necessitate corrective action being taken, and consequent non-steady-state behaviour.

Suppose, for example, stable parameters (some possibly defined in terms of rates of change) are operating in the environment of the firm and require the firm to allocate certain proportions of funds to various subsystems if steady state is to be achieved. Approximating the required allocations permits achievement and continuance of the steady state, while significant deviations from required allocations – say, an excess of resources allocated to marketing, or not enough to R & D – will result in firm/environment mismatch, and necessitate eventual corrective action. The firm which habitually mismatches resources and environment will exhibit erratic, unstable behaviour, while the firm which achieves a fair match between required and actual allocations would tend to exhibit steady-state behaviour.

Such characteristics of corporate decision-making do involve some potentially restrictive assumptions. Firstly, it requires that there is some regularity or pattern in subsystems with respect to their overall function and relations with their external and internal environments. This is implicit in the concept of 'Gestalt' which was earlier suggested to be a possible feature of resource preferences at top levels in the corporation. However, the existence of pattern and regularity at high levels of abstraction, involving technological change parameters in dynamic environments, is well illustrated by the wide applicability of the logistic curve to growth of technological capability or 'key parameters'.[6] Such growth patterns have been demonstrated for a number of technologies, the degree of abstraction being such that typically only a single characteristic such as capacity/speed/power is involved. With respect to the possibility that top management may perceive a global pattern in terms of environmental parameters and 'matching' internal resources, it is encouraging that abstraction demonstrates pattern in technological change, the latter representing perhaps the definitive characteristic of turbulent fields.

Secondly, it implies that the environment is sufficiently hostile and other firms sufficiently competitive to motivate steady-state-seeking behaviour. By competitive environment we simply mean that the environment is sufficiently unstable and demanding as to mitigate against significant discretionary allocations away from the steady state, and this appears to be guaranteed to a large extent by the pervasiveness of turbulent fields (Emery and Trist, 1965; Terreberry, 1968). In this respect, Williamson (1971) suggests that the modern M-form corporation is potentially more efficient in terms of profit-maximising than its predecessor, the U-form. As we saw in Chapter 4, the M-form facilitates separation of strategic and operating decisions, creating capacity at top-level decision-making for consideration of strategic change. It also reduces strictly functional matters to divisional level, with the consequence that partisan goal-seeking can be controlled or even eliminated; the top-level decision-makers have a 'psychological commitment' to the operation of the corporation as a whole (Williamson, 1971, p. 359). Williamson argues that as a consequence the M-form organisation and operation is less favourable for the indulgence and pursuit of managerial discretion than the alternative U-form owing to the 'superior efficiency, motivational and control properties' of the M-form (p. 367). Given the widespread adoption of this organisation form in modern U.S. industry, the implication is that the large corporation typically may face strong effective competition – whether real or potential – in its various markets. In such circumstances, we assume that corporate management seeks the steady-state solution that permits it to survive and also possibly to achieve some growth. We shall discuss this latter possibility further in a later section.

Thirdly, it implies limited substitutability of subsystems. If steady state could be maintained by replacing some R & D resources by equivalent expenditure on marketing, the steady state would not be a unique solution. In such circumstances, enduring stable preferences for resources may not reflect attainment of a unique steady state, but may reflect stagnation of resource preferences due to the adequacy of a range of alternatives. A problem considered below is how the framework may be specified to obtain unique allocations in the steady state.

THE DERIVATION OF THE PREFERENCE SYSTEM

In open system, the means whereby steady state is achieved is through feedback from its environment, particularly negative feedback: 'Negative feedback is informational input which indicates that the system is deviating from a prescribed course and should readjust to a new steady state. . . . Management is involved in interpreting and correcting for this information feedback' (Kast and Rosenzweig, 1974, p. 117).[7] Thus, as suggested earlier, third-level mechanisms contribute to the maintenance of steady state in an open system. However, such mechanisms do not provide information about steady-state characteristics as such; they are specified simply in terms of discrete action/inaction, negative feedback prompting corrective action, and absence of negative feedback signifying no system correction necessary. For derivation of preference systems leading to purposive steady-state-seeking behaviour, we have to formulate open system models of corporate behaviour. Negative feedback homeostatic mechanisms are useful for signalling when required steady-state allocations are being deviated from, and if the existing preference system requires adjustment. As the firm's experience and knowledge of a particular environment increases, so pattern and regularity in the features and performances of respective resources contributions is perceived. Morton (1967, p. 25) gives customer error, feedback through intermediate institutions, flow of funds from the external investment market, and articulation of new needs from market signals as examples of informational feedback in the Bell communications system. While some of this is error or negative feedback, some will also signal opportunities and successful strategies (positive feedback), and together both types of feedback may contribute to the elaboration of a stable preference system.

The previous section suggested that in an open system framework the managerial task is to achieve a 'match' between organisation and environment in terms of steady-state allocations to resources. Environmental parameters determine the preference system of the firm through the firm's interaction with the environment. It is not necessary that the various parameters be isolated and identified by the firm; it is the contribution of subsystem combinations to steady-state behaviour that is important. A Gestalt may be derived with respect to the pattern of resources that management have learnt to be effective from past experience of resource allocation in a particular environment. Although the specification and interrelationships of system and environmental parameters may be extremely complex, provided there is regularity and consistent pattern in system, environment, and their relationship, a stable preference system for resources may be built up through simple system–environmental interaction. Environmental turbulence need not imply lack of pattern; even though water movement in a rough sea may appear chaotic and irregular when viewed from close quarters, there may be discernible regular wave patterns when viewed from a greater height. Similarly, turbulence in corporate environment may be consistent with pattern at higher levels of abstraction and decision-making.

Once a 'Gestalt' appropriate to the system and environment has been achieved, the control problem may be reduced to a third-level problem, provided the pattern and configuration on which the Gestalt is based is expected to persist indefinitely; negative feedback would signify that the perceived pattern and corresponding resource allocation requires adjustment to changing circumstances.[8]

This is analogous to Friedman's example (1953, p. 21) of a billiards player who learns required actions and performs them skilfully without solving the complex and dynamic equations with which his shots may be represented in a mathematical model. An expert billiards player does not require knowledge of Newton's laws of motion, nor does he have to separate out relevant variables such as coefficient of friction, ball velocity, etc.; his skill will have been developed through the trial-and-error process of negative feedback. Similarly, management need not separately identify parameters and specify relationships for attainment of steady state, since this is learnt from experience of past allocations and consequent effects on attainment of steady state. Negative feedback adjusts the preference system and directs resource allocation to the steady-state value. Our feedback mechanism for the derivation of the preference system may be compared to problemistic search in behavioural theory, owing to the similarity between negative feedback and the satisficing principle. Given these similarities, a paper by Day (1967) is extremely useful in putting the relationship between derivation and specification of the preference system in our analysis in perspective.

Day argues that there is an 'underlying unity' between behavioural theory and neoclassical optimising theory. He demonstrates this by using a behavioural learning model in which the decision-maker is ignorant of his profit function and instead uses simple rules of thumb, based on the effect that changes in output have on profit. In this context there are two basic principles assumed: (1) repeat successful and avoid unsuccessful behaviour; (2) be more cautious if a repetition of previously unsuccessful behaviour appears desirable in future. Goal achievement is signified when changes in output lead to only a 'small change' in profit (p. 307). Using these and one or two other simple assumptions, Day demonstrates that the model converges on the neoclassical solution MC = MR.

The importance of Day's analysis lies *not* in demonstrating that differences in decision-making behaviour can lead to the same solution (e.g. that a 'behavioural' type of firm will allocate resources in a similar fashion to a 'neoclassical' type of firm), but rather that different models of decision-making can adequately represent the same behaviour, or different aspects of it. This optimising theory may model steady-state allocations, while behavioural theory (or related variants) may describe the process of attainment of that solution. In this respect at least they may be regarded as *complementary* rather than competing theories, an interpretation still not recognised by many economic investigators.

Therefore, models representing optimal decision-making can still be useful even in the absence of perfect knowledge (as was first argued some years ago by Machlup, 1946). The decision-maker may demonstrate optimal behaviour patterns without being able to articulate the reasons for his actions. Consequently, although the models developed here, and later in the analysis, are mainly based on optimising principles, they have essentially behavioural principles underpinning their main assumptions (e.g. use of the concepts of resource and feedback in preference system description and derivation respectively). Appreciating the relationship between the behavioural and optimising principles is essential to place the present analysis in context. Since we shall be looking at steady-state behaviour in the present analysis, our models will be specified in optimising terms generally. This still leaves open the option to use reactive behavioural models to describe non-steady-state behaviour, in effect the other side of the coin. Which group of models and assumptions should be used should depend on circumstances and the objectives of the analysis rather than dogma, and in this respect the building of the present analysis on a behavioural foundation may be regarded as a strength rather than as a weakness.

A FRAMEWORK FOR INTRA-FIRM ALLOCATION

We assume management makes subjectively rational decisions on the basis of a stable system of preference for resources, in the context of the corporation behaving as an open system in turbulent fields. Research and development may be an integral component in the preference system. In developing our framework these constitute the basic assumptions.

Since it has been suggested that the project-oriented neoclassical approach is inappropriate for analysis of corporate decision-making in turbulent environments, it might be expected that conventional economic theory in general is redundant as far as such analysis is concerned. In fact, given the basic assumptions above, this may not be the case. There exists in contemporary economics a well-developed methodology and substantial literature based on the assumptions of 'subjective rationality' and 'stable system of preferences' in the theory of consumer choice, particularly with respect to indifference curve theory. Concepts derived from indifference curve theory will be utilised in developing the framework.

We assume a firm divided into two distinct functions: production/marketing (P/M) and research and development (R & D). Utility is derivable from allocation of resources to both functions, perceived utility depending on the contribution a particular combination of resources makes towards the survival of the firm; we deal with possible components of utility later. In turbulent environments, survival is the prime objective of the firm and resources are deployed with this objective in mind. The more resources available to the firm, the better its chances of ensuring survival, *ceteris paribus*; however, the effectiveness of contributions of 'bundles' of resources (or combinations of subsystems) depends on the specific combination of resources employed over a defined period. The importance of the steady state in this context will be discussed later.

As far as the resource allocation to each function is concerned, it may be specified with respect to real expenditure on the resources employed in each function as long as relative prices are held constant; we shall assume that all firms have equal access to resources, and all resources are available to the firm at the market rate. If relative prices are assumed to be constant, P/M and R & D allocations may then be treated as involving homogeneous resources; this is analogous to Hicks' composite commodity theorem (1946, p. 33) except that we specify two composite resources instead of one composite commodity.[9]

The utility function may be defined as

$$U = f(Y_1, Y_2)$$

where Y_1 is R & D resources and Y_2 is P/M resources.

We shall therefore assume that our preference system deals with two resources, P/M and R & D, and that we can identify different bundles of resources, X, X', X'', . . . , X^n. A suggested set of assumptions permitting the development of a theory of resource choice is set out below, based closely on Green's axioms of choice (1971):

(1) Different allocations of P/M and R & D can be represented by a weak preference ordering; for any pair of resource allocations X and X', either XRX' or $X'RX$ or both (*completeness*) and if XRX' and $X'RX''$, then XRX'' (*transitivity*). The relation R means that one bundle is regarded as *at least as good as* the other.[10] If we obtain a function which assigns to each allocation a unique real number, then

$$U(X) \geqq U(X') \text{ if and only if } XRX'$$

i.e. the utility derived from X is greater than or equal to that derived from X' if and only if X is regarded as at least as good as X'.

(2) If X is a chosen allocation from a set of alternative allocations and X' is an allocation in that set, then XRX' (*rational choice*). In terms of utility, $U(X) \geq U(X')$ if and only if XRX'.

(3) If X contains at least as much of both resources and more of at least one resource, then X is preferred to X'; that is, if $X \geq X'$, then XPX' (*non-saturation*). If the utility function is differentiable everywhere in the set of alternative allocations, then the partial derivatives with respect to P/M and R & D are greater than zero.

(4) For any allocations X and X' a continuous series of allocations can

be found connecting X and X' (*connectedness*).

(5) The set of all possible allocations X' such that $X'RX$ is strictly convex for all possible allocations of X (*convexity*: the implications of this assumption will be discussed below).

(6) The marginal rate of substitution between all pairs of allocations in the set of alternative allocations is uniquely determined (*smooth indifference curves*).

One assumption that is both central and more complex than the others is the convexity assumption. We shall discuss this below, but first we shall assume that both functions exhibit diminishing marginal utility; as the use of resources in each function increases, so the contribution towards steady-state maintenance derivable from each function tends to decrease, i.e.

$$U_{11} < 0 \text{ (R \& D)}$$

$$U_{22} < 0 \text{ (P/M)}$$

The marginal utilities themselves, U_1 (R & D) and U_2 (P/M), are always positive (from the assumption of non-satiety). However, we are faced with a problem since our assumptions are not strong enough to guarantee a convex indifference curve, i.e. a diminishing marginal rate of substitution. Diminishing marginal utility is neither necessary nor sufficient for convexity (Green, 1971, pp. 83–94, 305–8). We can deal with this if we add the 'special hypothesis' of independent utilities (Green, 1971, pp. 89–94). However, there are strong reasons to suspect that the utilities dealt with here are definitely *not* independent, and sufficiently dependent on one another to affect the relevance of a model built on this assumption; the effectiveness of P/M will be affected by the amount of R & D supplying it with new ideas, and the effectiveness of R & D will depend on the amount of P/M available to exploit R & D ideas. We cannot assume that utilities are independent, therefore

$$U_{12} = U_{21} \neq 0 \text{ } (U_{12} = U_{21} \text{ from Young's Theorem})$$

However, it can be shown that indifference curves are strictly converse to the origin *if and only if*

$$U_{11}(U_2)^2 - 2U_{12}U_1U_2 + U_{22}(U_1)^2 \leq 0 \text{ (see Green, 1971, p. 90)}$$

We know that U_{11} and $U_{22} < 0$; U_1 and $U_2 > 0$ according to our assumptions. Therefore, for a convex indifference curve with a diminishing marginal rate of substitution, we require

$$U_{12} \geq 0 \text{ or } \frac{\partial^2 U}{\partial Y_1 \partial Y_2} \geq 0$$

This requirement means that, as we increase P/M resources, the utility derivable from additional R & D resources does not diminish. Normally we would expect the usefulness of R & D resources to increase as more P/M resources are made available to exploit the ideas and innovations generated by R & D, implying $U_{12} > 0$ and diminishing marginal rate of substitution. A possible qualification to this general conclusion is discussed in a later section of this chapter.

Unit measurement of composite resources may be defined independently in the respective R & D and P/M functions. Thus, for example, a given real expenditure may purchase *m* R & D resources or *n* P/M resources, where *m* may or may not equal *n* depending on measurement convention in the respective functions. As far as total expenditure on resources is concerned, managers face a budget constraint of expected available funds for resource allocation to be distributed according to the resource preference system. Given the constraint of available system expenditure, allocations to subsystems can only be increased by trading off allocations to competing subsystems.

The utility of a resource combination is specified in terms of contribution to the survival of the firm; in a benign environment long-term survival may be consistent with a variety of alternative allocations, but we assume this is not the case here. Corporate allocations are oriented towards ensuring the survival of the firm in a dynamic, hostile environment; attainment of a steady-state allocation which may be perpetuated indefinitely is an essential part of the problem of maintaining organisational viability in turbulent fields. Maximising utility of resource allocations therefore involves attempting to establish the particular steady-state solution which will ensure organisational viability. In general, it should be possible to identify the possible elements of utility derivable from a given resource bundle: R & D as a source of innovations, marketing for selling effectiveness, etc. Ultimately the utility of functions must be seen in terms of their capacity for revenue generation in conjunction with other functions in the firm, though *not* at project level generally. We assume that it is possible to establish a pattern or overview in terms of revenue generation capacity of specific bundles of resources.

Obviously, since the payoff from allocation to R & D resources is typically highly lagged, the time period over which the preference system is hypothesised to operate must be long enough to permit most viable projects to be worked through to development and subsequent exploitation. Note that this does not necessarily mean that the utility derivable from R & D resources lags, the utility from production for any time period; as emphasised, utility is derivable from *combination* of resources, R & D requiring eventual use of P/M resources before utility can be generated by any R & D allocations. It is the balance of resources or combination of subsystems which constitutes the prime decision-making problem at this level.

The allocation problem can be described graphically (see Fig. 5.1). For a resource expenditure constraint of aa' management may allocate a maximum of $0a$ resources to R & D and $0a'$ to P/M. Management are motivated to allocate resources intended to achieve steady state owing to

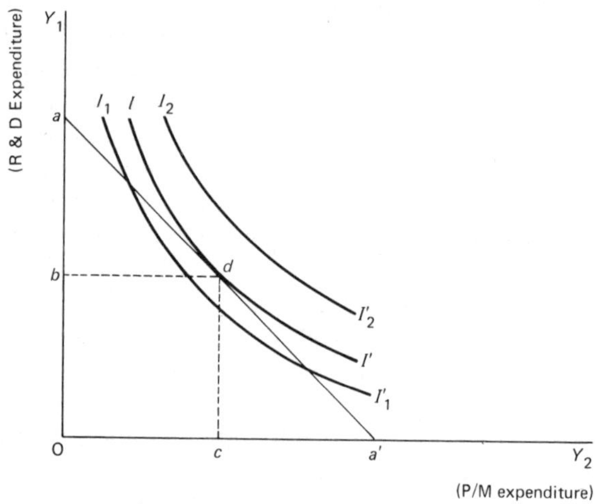

FIG. 5.1.

the environmental pressure in competitive, turbulent fields and consequently attempt to maximise the perceived utility of resource contribution towards the steady state. It is assumed that the demands of the environment are such as to not permit any correlate of organisational slack in corporate resources, and that the steady state is a unique solution with respect to the specific level of corporate resource expenditure incurred. The implications of this assumption are discussed further in a later section.

The indifference curves in Fig. 5.1 describe the locus of points providing equal perceived opportunity for long-term survival. All resource combinations described by a particular indifference curve are seen as providing the same potential in terms of the present value of operations conducted over the specified time period, even though the components of utility (such as advertising campaigns, inventions, etc.), and the manner in which future revenue is generated, are liable to be different for each resource combination.

The implication for management described by Fig. 5.1 is that, given the resource budget constraint aa' and the managerial utility function, utility is maximised at d where the highest possible indifference curve II' is attained. II' is tangential to aa' at d. Allocations on, say, I_1I_1' result in a resource mismatch between firm and environment; combinations on higher indifference curves are preferred as allocations approach the steady-state solution, and unless the steady-state solution is achieved, in the long run the corporation will be non-viable. On the other hand, allocations on I_2I_2' are too expensive for the firm to consider even though they generate greater utility than I_1I_1' combinations. Point d is therefore the highest available utility point attainable with available resources, and since the preference system is expressed deterministically, corporate management are assumed to have knowledge of this combination of resources, the combination most likely to guarantee future survival in the face of actual or potential competition; it is assumed that at this level of abstraction there is no significant uncertainty pertaining to resource allocation.

In short, given the resource budget described by aa', attainable allocations other than at aa' are inefficient in that for the available resource expenditure they provide less than maximum expected utility. Since efficiency in the sense of utility maximising is assumed to be necessary for long-term survival owing to environmental pressure, for resource budget aa', allocations other than at d would be non-steady-state allocations, and not tenable indefinitely.

Utility maximisation does not guarantee attainment of the steady state. Firstly, managerial preferences may be incorrectly specified for steady-state behaviour. Negative feedback from performance of resources should adjust preferences towards the appropriate steady-state system of preferences. Overly slow adjustment of preferences, or persistently incorrect preferences, will result in eventual failure of the firm. Secondly, it may be that steady-state is not feasible for a particular firm in a particular environment. Unless firm or environment is changed sufficiently rapidly, it will result in the failure of the firm. We shall assume that 'natural selection' 'weeds out' firms which cannot attain a steady-state match with their environment. 'Natural selection' necessitates that stable preference systems for resources, coincidental with those required for steady-state behaviour, evolve through feedback from the environment and result in firms capable of developing appropriate and effective steady-state preference systems. We therefore assume that surviving firms are both able and required to attain steady-state allocations intended to maximise system viability.[11] This 'survivor' hypothesis alone, however, is not sufficient to guarantee utility-maximising behaviour; as Alchian (1950) points out in an early statement of the natural selection argument in an economic context, 'even in a world of stupid men there would be profits'. Natural selection favours the *relatively* efficient, and there is no requirement that these are profit or utility maximisers. In order to justify the assumption of utility maximising, natural selection would appear not to be sufficient.

In fact, the assumption is more reasonable in this analysis than in Alchian's, since the preference system is deterministic in the present case, while in Alchian's analysis firms operate in conditions of pervasive uncertainty. Because of the level of abstraction and application of the preference system, the firm is assumed to have adequate knowledge of the steady-state balance of subsystems, and believes that if this steady state is not attained and maintained, it is vulnerable to attack from other corporations who may perceive that it is mismatched with its environment. The relevance of uncertainty in our analysis is restricted to lower levels and micro-processes in the corporation, and consequently the assumption that corporations are utility-maximising steady-state seekers is facilitated by the assumption of determinism at higher levels of abstraction.

It is important to establish how such utility functions may be related to corporate budgetary conventions discussed earlier. It has been suggested that the indifference curve analysis must be defined over time periods long enough for the utility derivable from R & D to be generated, in effect the very long run when technological change can be incorporated into firms' production functions and product lines. Yet usually corporations decide to allocate funds to R & D on an annual basis, typically using the rule-of-thumb techniques described earlier. Is it possible to reconcile differences between the utility analysis and budgetary conventions, and if so, how?

It must first be emphasised that the choice of time period for a particular set of indifference curves and resource constraints is arbitrary, provided of course that they extend into the very long run. This is because the basis on which utility is calculated is the same for all such indifference curves, irrespective of actual time periods over which they are defined; expected utility of future subsystems allocations is calculated from feedback on *past* performance and derived utility of all previous allocations up to the present time. Thus, in Fig. 5.1, the actual time periods over which the indifference curves and resource constraints may be defined could be, say, fifteen or thirty years. Essentially there would be no difference between the two different sets of indifference curves and resource constraints, except that in the latter case utility and resource constraints would be magnified by some factor compared to the former case, e.g. a doubling in the case of expected zero growth of the corporation.[12]

Since we have assumed that the corporate management seek a steady state, the gross allocations perceived to be required for survival in the long run may be annually distributed to subsystems according to Von Bertalanffy's definition of the steady state as a time-independent state in which the ratio of the system's components is constant (1973, p. 167). However, this does not mean that the same absolute resource expenditure may be distributed to subsystems in each time period; even though we assume that the primary objective of corporate allocations is survival, this may be consistent with regular steady-state expansion of the resources budget. Corporate growth in an expanding industry may be a normal part of corporate operations; the static firm in such circumstances may be seen as potentially vulnerable and weak, inviting attention from predatory rivals. Thus, pursuit of the primary corporate objective of survival may be reflected in some degree of corporate growth in subsequent time periods.

If we assume that the particular set of indifference curves and resource constraints in Fig. 5.1 is defined over n years, and that the expected growth of the corporation in terms of available resource expenditure per annum is h per cent, then the conditions we assumed for steady-state expansion of the firm measured on an annual basis are:

$$X_1 + \left(\frac{100+h}{100}\right)X_1 + \left(\frac{100+h}{100}\right)^2 X_1 + \ldots + \left(\frac{100+h}{100}\right)^n X_1 = 0b$$

$$X_2 + \left(\frac{100+h}{100}\right)X_2 + \left(\frac{100+h}{100}\right)^2 X_2 + \ldots + \left(\frac{100+h}{100}\right)^n X_2 = ba$$

where X_1 and X_2 are the average resource expenditure in the first year for Y_1 and Y_2 resources respectively. Resource expenditures $0b$ and ba are measured in Y_1 units (see Fig. 5.1). The above states that if the firm is expected to grow in terms of a particular annual growth rate, then the corporate subsystems will be expanded at the same annual rate over the period for which the indifference curves and resource constraints are defined. This not only satisfies the Von Bertalanffy definition of steady state, it is also consistent with the widespread corporate convention, discussed earlier, of annually allocating funds to R & D and other functions on a percentage of sales basis. In the simple case where zero growth of available corporate resources is expected, the original annual allocations to subsystems persist throughout the time period and $X_1 = 0b/n$, $X_2 = ba/n$. This factorisation of the components of utility into constituent annual resource allocations according to steady-state rules is one interpretation which may account for the rule-of-thumb percentage-of-sales budgeting practices, though of course other possible explanations could be suggested.

As far as the *determinants* of resource allocation are concerned, Von Bertalanffy has shown that, for open systems, time-independent steady states may depend in certain circumstances only on system-specific characteristics. We shall similarly assume that there exists for each firm a unique steady-state solution determined in part by *m* intra-firm characteristics, $\phi_1, \ldots, \phi_m$; these intra-firm characteristics may vary for different firms in the same industry, resulting in a variety of steady-state solutions:

$$V_i = f_1(\phi_1 \ldots . \phi_m)$$

where V_i is the steady-state solution defined in terms of fraction of available resources allocated to a specific subsystem for the *i*th firm.

Von Bertalanffy's analysis of steady-state determination emphasises systems' characteristics rather than environmental characteristics: provided the environment of the system generates sufficient life-maintaining inputs, the environments may be regarded as homogeneous to all intents and purposes for intra-species analysis if mechanisms convert the various inputs into homogeneous units through various species-specific energy conversion exchanges. Emery and Trist (1965) and Ashby (1960) have shown, however, that steady-state behaviour may depend on environmental characteristics, especially if the system under consideration is a social system which does *not* have structurally determined input conversion techniques. The system in such circumstances adapts to its environment and searches for a steady-state solution appropriate to, and partly determined by, the environment. If it is assumed that the industry or industries in which a firm operates are synonymous with its environment, we can extend the above equation to cover environmental characteristics:

$$V_i = f_2\ (\phi_1, \ldots, \phi_m, I_1, \ldots, I_n)$$

where $I_1, \ldots, I_n$ are industry characteristics for the industry in which the ith firm operates.

A further definitive quality of open system is implicit in the model above, that of equifinality. According to Von Bertalanffy, 'If open systems . . . attain steady state, this has a value equifinal or independent of initial conditions' (1973, p. 140).[13] The final state is determined by the system and environmental parameters and relations; consequently, even if system actions and allocations bear no relation to the steady state originally, it will move towards the steady-state solution which exists *independently* of existing allocations. Otherwise the system will eventually fail. A corollary is that if two or more systems (firms) are operating under similar system and environmental parameters, the steady-state solution will be equivalent for all systems (firms).

Thus, our systems model is specified as a utility-maximising deterministic model with static preferences for resources. While such a formulation may appear extremely restrictive, it will be justified if the evidence considered in the subsequent chapters is consistent with such description of corporate behaviour. One aspect which should be emphasised, however, is the restricted class of decision which will be studied. We are concerned with allocations to functions decided at high levels of abstraction, and only steady-state allocations in this subset of corporate decisions. As suggested earlier, behavioural theory might be appropriate for dealing with non-steady-state behaviour, models of the type suggested above complementing rather than substituting behavioural theory.

UNIQUENESS OF THE STEADY STATE

It is a basic assumption of the model developed above that there is a single steady-state allocation which firms must attain if they are to survive in the long run. However, there are possible objections to the restrictiveness of this assumption and two of the most potentially significant are discussed below.

Firstly, there may be a substantial threshold as far as the minimum size of R & D laboratory is concerned, violating the assumption of strict convexity. In such circumstances there may have to be an explicit choice between research-intensive operation and virtually no interest in R & D, as far as the individual firm is concerned. There may still be room for firms adopting the latter course if they adopt a 'subordinate' or 'dependent' strategy (see Freeman, 1974, pp. 274–6) in which they act as subcontractors for large firms in the industry. The large firms may tolerate the existence of smaller firms operating in this fashion since dependent firms may act as 'buffers', taking up much of the variation created by environmental turbulence, and survival may also be facilitated by low overheads or other special advantages at least partly offsetting their strategic disadvantage.

Since 'dependent' firms tend to be small, satellite firms, the potential significance of the above qualification is diminished to some extent by the emphasis in this book on the *large* corporation, both in the preceding analysis and in the studies of the next three chapters. The possibility of threshold effects in certain technologies being large enough to have a significant effect on corporate decision-making even in large corporations must, however, be recognised.

Secondly, it may be the case that competition is not strong enough to ensure a unique solution in terms of allocation of resources. This is particularly likely to be the case if there exist barriers to entry or competition in particular industries, and in such circumstances managerial discretion over the pursuit of goals other than the survival of the firm may be permitted. To the extent that competition is impeded and managerial discretion over resource combination is feasible, there may be more than one resource allocation solution consistent with survival. Further, the higher the substitutability of subsystems, the less sharply peaked will be indifference curves with respect to specific budget constraints, and the greater the range of feasible managerial discretion. This is demonstrated in Fig. 5.2, in which indifference curve aa' implies a high degree of subsystem substitutability, while bb' implies a low degree of such substitutability. In both cases, combination X represents the unique combination necessary for survival if $0m$ resources are available (measured in Y_1 units). However, if there is available resource expenditure in excess of the minimum required for survival, of mn (in Y_1 units), then the range of discretion over resource combination is ps in the case of high substitutability, but only qr in the case of low substitutability. Thus, a small amount of slack resources may be consistent with a relatively wide range of managerial discretion if there is a high degree of subsystem substitutibality. As was implied earlier, substitutability of

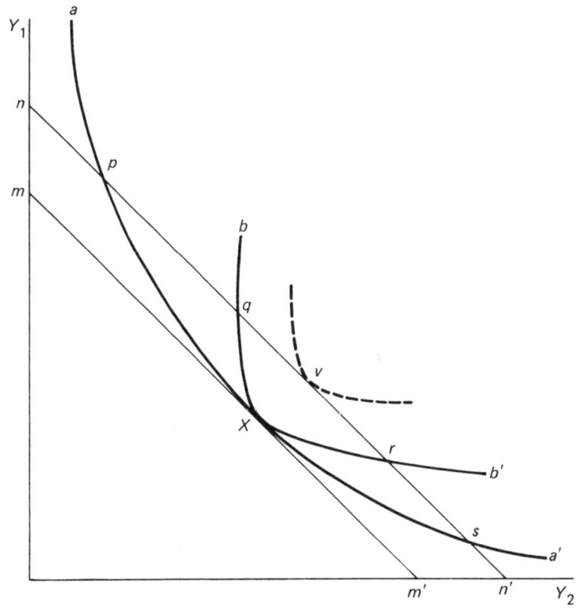

FIG. 5.2.

subsystems may threaten the uniqueness of the steady state and, as the above demonstrates, degree of substitutability is closely related to the range of permissible managerial discretion.

However, there are at least two qualifications to the possible importance of managerial discretion in the present analysis. Firstly, barriers to entry typically refer to individual products and processes, e.g. economies of scale, patent protection, advertising, etc. It is worth re-emphasising that our concern here is at *functional* level in the corporation, and that while specific barriers may exist at lower levels, the significance of a particular barrier to entry is liable to be diminished at higher levels unless a corporation has an across-the-board advantage which cannot be nullified or replicated by competitors working in other product areas. An example of this might be a highly productive R & D team whose creativity and inventiveness cannot be matched by competitors.

Secondly, even if a corporation appears to have nominal discretion over the allocation of funds to subsystems owing to relative weakness of existing competition, there may still be factors inducing resource utility maximisation. The threat from *potential* competitors may encourage utility maximisation and dissuade management from pursuance of other goals, in case they lose ground to new entrants who perceive that management are not taking advantage of all their opportunities. Also, fear of take-over could stimulate resource utility maximisation if failure to do so signalled vulnerability to potential raiders. Both these possibilities would encourage maintenance of resource utility maximisation as a means to long-run survival, which in the case discussed above would mean utility with available resources $0n$ (measured in Y_1 units). This is achieved at point v on the dotted indifference curve in Fig. 5.2.

In practice, firms are liable to achieve some degree of discretion at various points in their history, for long or short periods of time. In such circumstances, management may pursue non-survival-oriented goals, such as sales, emoluments, staff employment, etc. It is to be hoped that in the present analysis these possibilities do not distort resource allocations sufficiently to render impracticable the testing of hypotheses based on the framework developed here.

One last point should be emphasised, however. Uniqueness of steady-state solutions for individual firms does not necessarily imply it will be the same for all firms in a particular industry. As suggested earlier, the steady-state solution is determined by the interaction of intra-firm and environmental variables, and consequently differences in the specification of individual firms in an industry may require differences in the R & D profile or budget.

HIERARCHICAL STRUCTURE OF PREFERENCE SYSTEM

The model above is concerned with resource allocation between two functions, R & D and P/M. However, it is based on the idea that there are identifiable systemic properties at high levels of abstraction in the overall system. Allocation is top-down, higher-level parameters and relations determining preferences at a particular level, which in turn provides the expenditure available for distribution among lower-level systems. To the extent that stable patterns are perceived at the lower level, preference systems for resources with respect to subsystem allocations may be operational at those levels. Rather than simple aggregative 'bottom-up' allocation, there may be a hierarchy of allocations of resources; allocations to systems at higher levels may constrain allocations to systems at lower levels. For example, in Fig. 5.1 the allocation of $0b$ resources to R & D constitutes a constraint for all allocations to R & D subsystems. The lower down the hierarchy subsystems operate, the more unstable we would expect subsystem preferences to be; this is analogous to the inferred stability of 'meta-rules' referred to by Nelson (1972). Pattern perceived at higher levels may not be paralleled by stable 'Gestalt' at lower levels where environmental turbulence would be expected to create subsystem vulnerability.

However, assuming that stable preference systems exist at subsystem levels in the intra-firm allocative process, it is a natural development of the above model that a hierarchical arrangement of system preferences may be identified, lower subsystems being constrained by resources allocated from higher subsystems. This is illustrated in Fig. 5.3, which

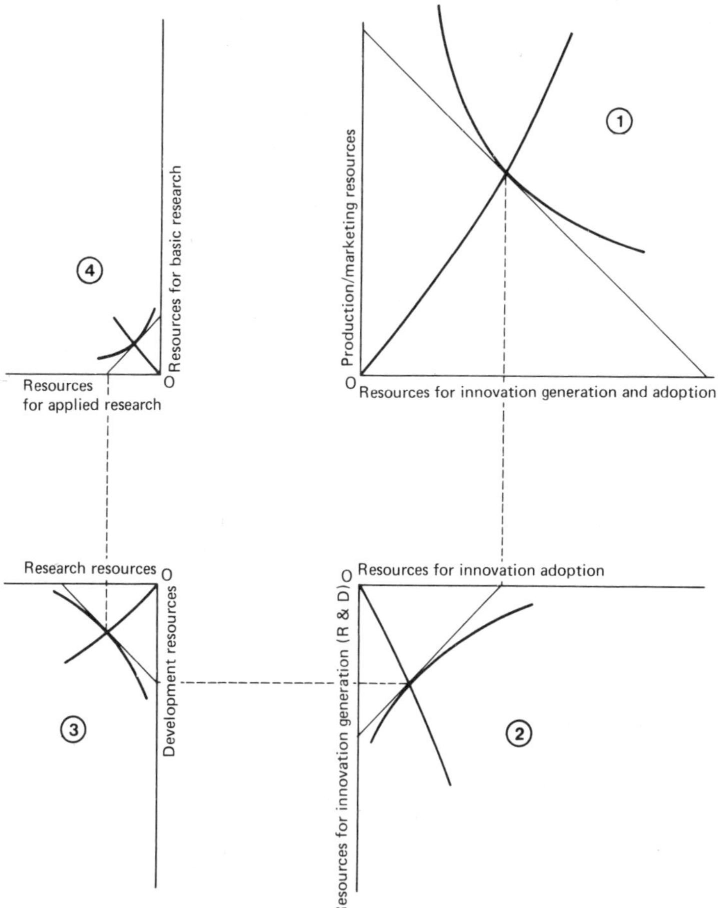

Fig. 5.3.

shows a natural hierarchical framework within which resource preferences might be arranged. Choice is made between resources for maintenance subsystems (production and marketing) and adaptive subsystems, in quadrant 1. Resources allocated to this latter category constrain resources available for allocation to search for externally generated innovation and R & D (see quadrant 2). The chosen resource allocation for R & D in quadrant 2 constrains resources to be distributed between the development subsystem and research subsystem (quadrant 3). Research funds in turn are allocated between basic and applied research according to the preference system in quadrant 4.

It is assumed that there is only a simple two-subsystem choice to be made at each system level in the hierarchy. More complex decisions may be conceived, e.g. a simultaneous three-subsystem decision as to whether to allocate R & D funds to basic research, applied research or development, and indeed the problem may be made more general by analysing it in terms of constrained allocations to *n* subsystems. However, the 'top-down' hierarchical arrangement of allocations simplifies the problems of analysis facing the decision-maker at each particular level, and the evidence of the last chapter suggests that this restriction is employed in actual decision-making (e.g. the simple two- or three-subsystem decision where resources are allocated to R & D and production/marketing). The actual subsystems involved and their place in the overall system hierarchy may of course depend on circumstances.

For the moment, this outline of resource allocation decision-making will suffice to provide a basic model of corporate decision-making. A modification will be introduced in Chapter 7 to deal with cases where the assumption of strict convexity is violated; some R & D performing firms do not conduct basic research, while many firms do not conduct any R & D in the first place. In Chapter 4 it was suggested that firms tended to prefer less uncertain subsystems to more uncertain systems, *ceteris paribus*, and we will subsequently investigate circumstances in which certain subsystem activity is not undertaken, as well as the factors encouraging take-up of previously neglected subsystem activity.

CONCLUSION

The purpose of this chapter has been to develop a holistic and hierarchical framework of functional allocations in the firm, based as closely as possible on the observations of business behaviour and related open-system interpretation of Chapter 4. A number of related and mutually reinforcing threads facilitated this development; the apparent existence of Gestalt or pattern at high levels of abstraction, the relevance of open-system concepts such as differentiation and steady state to corporate behaviour, and the observed tendency for allocations to diffuse from higher levels to lower levels, have all contributed to the development of a systemic, non-aggregative model of the firm.

As Rapoport and Horvath (1959) point out (attributing the original insight to Alfred North Whitehead), the 'constraining framework of thought' in the development of science has been the overwhelming emphasis on analytic thinking, that is, in attempts to understand complexity by examination of constituent parts. Summativity and aggregation of parts does in fact work in a wide range of observed behaviour, and as Rapoport and Horvath point out it leads to the temptation to generalise analytic thinking to all complex phenomena.

Analytic thinking has similarly dominated economic theory in its development as a social science, and it is only recently that non-divisible concepts at a higher level than the project have entered into economic theory, for example behavioural and Penrosian theory. In fact, the significance of such development in the case of behavioural theory was generally overlooked, attention being concentrated on the debate as to whether or not it constituted a genuine, refutable 'theory', or to what extent it could generalise to different organisations. In this chapter we have attempted to develop a hierarchic, holistic approach to corporate decision-making in which pattern or 'Gestalt' formation replaces the reductionist perspective implicit in neoclassical theory.

The simple systemic model of the firm developed in this chapter will provide a basis for study of observed corporate attitudes and allocations to R & D. It will first be utilised in attempting to explain differences in decision rules for allocation of funds to the R & D function in Chapter 6.

PART II

6 Budget Decision-Making for Research and Development

As an initial examination of the relevance of the systems frame of reference developed in the preceding chapter we shall examine the pattern and distribution of decision-making techniques employed by various types of firms in deciding the R & D budget. As will be demonstrated, there are occasionally substantial differences between groups of firms separated by time and international boundaries, and reasons will be suggested for such differences. The analysis of the chapter may be seen as a direct development of the early section of the previous chapter in which were discussed problems of abstraction in complex systems. It may be regarded as interpreting Simon's statement, 'Man . . . is a pattern finding animal', as a testable hypothesis since it discusses requirements for establishment of patterns in R & D budgeting, and attempts to explain differences in budgeting convention in this light.

It was argued in Chapter 4 that large modern corporations in the United Kingdom and the United States exhibit stable preferences for employment of R & D resources relative to other functions. The tendency for R & D budgets to be decided on a percentage of sales basis was cited as supporting evidence. This convention has been widely accepted as describing the main decision-making criteria adopted by such firms, as well as being frequently criticised for its 'unsound' or 'irrational' nature. Mention was also made in Chapter 4 that exceptions existed to the rule-of-thumb convention, and it is with these in particular that this chapter is concerned.

The first exception is provided by a recent survey of R & D budgeting in large Swedish corporations in research-intensive industries (Naslund and Sellstedt, 1974). It provides evidence apparently contradicting previous studies reporting widespread adoption of percentage of sales criteria. Of the 94 firms responding, 64 employed R & D resources in conducting their business; 'analysis of single projects'[1] was used 56 times as a method for determining the budget, 'per cent of forecasted or previous sales' 16 times, other methods 19 times.[2] Naslund and Sellstedt conclude that their study does not confirm previous studies and opinions emphasising the importance of rule-of-thumb methods (p. 70) and state that 'one underlying assumption in the work in this area seems to be that first resources are made available and then research and product ideas are produced. What we find indicates that it may sometimes be the other way around. It is the quality of ideas that explains the size of the budget. Good ideas might also generate their own funds making banks more willing to supply loans at favourable terms' (pp. 70–1). Näslund and Sellstedt also criticise the 'inflexibility' of the percentage of sales criteria (p. 69) and suggest that the role played by the R & D budget in explaining R & D success is vague and little known.

The project-based method was also reported in the United States by Bloom (1951) on the basis of a series of interviews with executives: 'The most common reply is that the number of worthwhile ideas added up to that amount [the budget]. Apparently research budgets are built up from the bottom rather than from the top down. Only in a few companies is a certain amount set aside for research – say, a certain percentage of sales – and then allocated among various projects. In most cases the ideas come first and then the allocation of funds' (p. 610).[3]

As with Näslund and Sellstedt's analysis, the quality of ideas and project-based budgeting is emphasised. Gambo (1959), writing a short time later, also contended that in the United States, 'in preparing a budget, a research program is first prepared by projects. Manpower is then allocated to the various projects and average salary rates are used to calculate the total salary budget' (p. 39). Both Bloom and Gambo base their conclusions on casual observation of industrial behaviour, and to that extent must be treated with reservations; Gambo in particular conflicts with the earlier description of R & D budgeting in a large firm by Reeves (1958) in which the R & D budget was established by a 'top-down' procedure.

There is at least one other area in which project-based budgeting for R & D has been well documented recently, that of *small* R & D performing firms in the United States. Hogan and Chirichiello (1974) and Smith and Creamer (1968) both conducted analyses of small company R & D budget setting, 'small' being interpreted in both cases as firms employing fewer than 1000 people. Hogen and Chirichiello, whose analysis is based on an N.S.F. survey and personal interviews with top officials, state: 'In the small company the R & D budget is usually "built up" by management's evaluating the proposals of each of the R & D professionals. These proposals are weighted against corporate objectives and available resources' (p. 28).

In another study based on a number of case studies of R & D performing firms in U.S. industry, Smith and Creamer (1968) conclude that as far as small firms in the machinery, chemical and electrical industries are concerned, 'While most of these companies do . . . provide for some form of budgeting, planning and review of technical programs at the corporate level, in no case was the person interviewed able to state a definite rule governing the level of annual expenditures' (p. 140).[4] This is in contradistinction to the evidence reviewed earlier with respect to budgeting in the larger firm, in which it was the consensus that percentage of sales constituted a definite and widespread method of budgeting; however, it would be consistent with the findings of Bloom, and Hogan and Chirichiello, in which the budget was revised not according to a pre-programmed method, but according to the situation as it prevailed at the end. Circumstances and opportunities appear to exert powerful influences over budget setting in the small firm.

REQUIREMENTS FOR ESTABLISHMENT OF STABLE PREFERENCES FOR R & D RESOURCES

Thus, studies in three separate areas emphasise that project-based approaches tend to be utilised, i.e. Swedish firms and small U.S. firms in the present day, and U.S. firms of the immediate post-war period. In contrast, studies of large U.S. and U.K. firms tend to suggest that the budget is decided from the top down. While noting the variability in method and reliability of the different studies, it would be a useful application of the model developed in the previous chapters if it could account for the differences between the two sets of studies.

In this respect, a significant feature of the hierarchic open system model is that it reverses the emphasis of neoclassical theory. In the former, consideration of system behaviour is based on a holistic concept of system separable from component elements, while in the latter the characteristics and behaviour of the system are totally determined by the properties of the consistent elements. A direct consequence of this reversal is that , *whereas in the neoclassical frame of reference it was the project-based budgeting methods that were consistent with the expectations of the model, in the systems model it is the budgeting techniques of the large U.S. and U.K. firms.* Consequently, in this frame of reference it is not the behaviour of firms employing budgetary conventions such as percentage of sales that are interpreted as deviants, but the other group of firms employing project-based methods. It is the aberrant behaviour indicated by the studies conducted in those areas which must be accounted for by the systems explanation.

To achieve this, it is useful to consider the circumstances in which one would expect the system frame of reference to hold. In Chapter 5 it was suggested that the corporation management perceive pattern in the systems and environment relationships and that perception of pattern accrues through managing and allocating resources in the firm. Appreciation of the capability and characteristics of different resource allocations develops through a learning process in which experience plays a crucial role.

Implicit in this interpretation are the assumptions that corporation management has sufficient experience of subsystem characteristics and linkages to establish an 'overview' of its situation and system/environment relations, and that individual projects are not significant enough in the content of the overall R & D function to affect managerial steady-state resource preferences. These assumptions may be formalised as two conditions for stable preference for resources:

(i) Management has extensive experience of R & D resources.
(ii) Component projects do not dominate the R & D programme.

Violation of either of these conditions would mean that a stable preference system could not be established. With this in mind, the operating conditions of the groups of firms predominantly employing project-based budgetary methods are discussed below.

(a) *U.S. firms in the late 1940s*

This period just precedes the widespread institutionalisation of research in U.S. industry, as Schon earlier pointed out. Although institutionalisation had been proceeding for some years, it is only in the later post-war period that R & D developed into an accepted, integrated function in the corporation and became 'professionalised'. Clare (1963) puts this development in perspective by indicating that the number of personnel employed in U.S. industrial R & D laboratories roughly doubled every $6\frac{1}{2}$ years from 1920 to 1959, and from 1945 to 1957 personnel employed quadrupled (p. 137). A further indication of the change in R & D relative to other resource uses is given in Fig. 6.1, in which is shown the growth of industrial research as a percentage of the

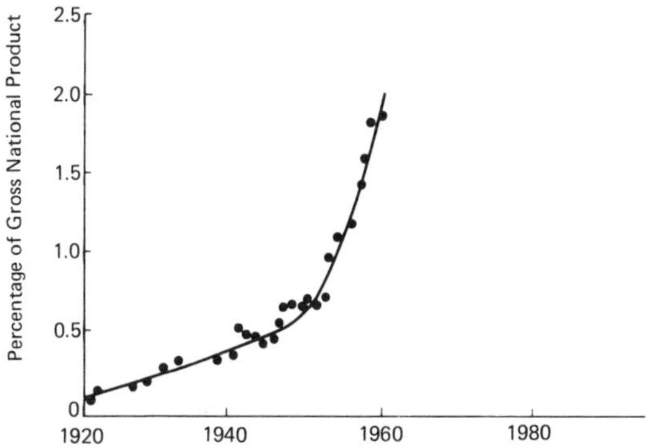

FIG. 6.1. Percentage of industrial research in the United States

Source: J. W. H. Clare, Current trends in the organisation of industrial research. *Research Management*, 6, 1963.

gross national product in the same period (Clare, 1963, p. 137). Figure 6.1 indicates that the growth of R & D relative to other resource uses increased rapidly in the 1950s after a period of relatively steady growth (apart from the 1940–5 period).

In such circumstances, corporation management was typically inexperienced in handling and integrating the R & D function – inevitably in conditions when the number of companies utilising R & D increased by 58 per cent in a five-year period (1950–5) (Clare, 1963, p. 137). Consequently, with R & D still a young and relatively unfamiliar activity to most firms, condition (i) above would not be generally applicable.[5] We would not expect stable resource preference systems to be generally operative.

(b) *Small U.S. firms in the present day*

The vulnerability of both requisite conditions is indicated when the operating conditions of small and large firms are compared. *Ceteris paribus* the larger firm allocates a greater absolute amount to functions compared to a smaller firm of similar R & D intensity and consequently there is greater opportunity to observe the characteristics of respective resources. Similarly, for projects of a given size and expenditure, condition (ii) becomes tenuous as size of firm decreases. Small firms tend to compensate for this at least partially by adopting low-uncertainty, short-term projects (Hogan and Chirichiello, 1974, p. 29), but if any significant degree of radicalness is associated with the small firm's R & D portfolio, the firm is liable to find absence of both conditions inhibiting the development of a stable resource preference system. We would therefore expect small firms to be less likely than large to exhibit stable preference systems.

(c) *Swedish firms in the present day*

Once it is pointed out that large Swedish firms are small relative to large U.S. and U.K. corporations, it is possible to place the Näslund and Sellstedt study in an appropriate perspective. Näslund and Sellstedt's sample of 'large' Swedish firms was constructed on a stratified basis of 20 per cent from firms with 100–199 employees, 20 per cent from firms with 200–499 employees, and 60 per cent from firms with over 500 employees. Further information on number of employees is not provided, but it would be reasonable to assume that the median firm does not employ significantly more than 500 employees. In the context of Swedish industry these are 'large' firms, as Näslund and Sellstedt's abstract suggests (1974, p. 67). However, not only is Sweden a smaller country with a smaller G.N.P. than the United States or the United Kingdom, it is also less research-intensive; O.E.C.D. statistics indicate that in surveys of R & D expenditures in each country in a three-year period (1963–5), in specific years R & D financed in industry as a percentage of net industrial output was 1.8, 2 and 3.3 per cent in Sweden, the United Kingdom and the United States respectively (O.E.C.D., 1971, p. 114). When government financing of industrial R & D is taken into account, the gap widens to 2.4, 3.2 and 7 per cent respectively. In such circumstances, what would be a large R & D programme in Swedish industry would be relatively insignificant in the other two countries.

In Näslund and Sellstedt's sample, 40 per cent of firms employed fewer than 500 people. In the United States in 1970, 82 per cent of funds for industrial research and development were distributed in firms employing more than 10,000 personnel (N.S.F., 1973b, p. 30). 'Small firms' in a U.S. context, as indicated by the Hogan and Chirichiello and Smith and Creamer definitions, would be a term applicable to most of the firms in the Näslund and Sellstedt sample. This interpretation is reinforced by Näslund and Sellstedt's finding that the president or board of directors usually decided whether to implement individual projects (p. 71). This facilitated the decision making-process of U.S. firms of the early twentieth century (see Chapter 4) and consequently top managerial decision-making is typically conducted at a much lower level of abstraction than in the large multi-divisional firms of the United States and the United Kingdom. Consequently the same arguments against the existence of stable preferences in small U.S. firms would be applicable to modern Swedish firms.

A further reason why Swedish firms tend to employ a project base for budgeting R & D may be the orientation of its research and development effort. Dörfer (1974) points out that the main Swedish R & D effort had been in the realm of 'big science',[6] defence and atomic energy (p. 141), with strong relations and spin-off between industry and government research in both directions (p. 138). By definition, big science involves large-scale projects requiring large commitments of funds, further weakening condition (ii) above in a context where firms have relatively small R & D programmes. Industry might also be expected to be influenced by project evaluation techniques employed by government when relations between industry and government are so close.

IMPLICATIONS OF STABLE PREFERENCES FOR R & D IN THE LARGE FIRM

In the previous section it was suggested that if either of conditions (i) or (ii) were violated, stable preference systems for resources would not evolve. The circumstances in which corporation groupings (a), (b) and (c) operate are such as to indicate the relative vulnerability of conditions (i) and (ii) in groups (b) and (c) and condition (i) in group (a).[7] Consequently, the exceptions (a), (b) and (c) usefully test the systems interpretation in being consistent with circumstances in which perception of stable pattern in resource allocation would be expected to be less likely relative to the large U.S. and U.K. corporations of the present day.

This encourages a re-evaluation of the R & D decision-making process in the light of the open system interpretation. The differences in budgeting techniques may be attributed to the differing abilities of firms to perceive stable patterns in relations between corporation resources and environment. In conditions where such pattern cannot be established or is easily disrupted, we would expect corporate allocation of resources to R & D to be based on vague, circumstantial, situational and/or individualistic techniques. This is in fact the type of behaviour typically found in the studies of groups (a), (b) and (c). The evolution of stable preference systems on a widespread basis is found in the large modern corporation, but is not apparently a common feature of the other studies reviewed.

This development of the arguments of Chapter 5 appears to account for differences in R & D budgeting convention between different groups of firms. There does not appear to be a corresponding explanation provided by project-based approaches such as the neoclassical theory of the firm. Of course this does not mean that no decision as to allocation between projects is taken, only that budgeting for R & D typically *precedes* allocation to projects in large corporations, as we would expect from Chapter 4. Our view of the firm is of a hierarchical arrangement of systems and component subsystems, disbursement of funds being from higher levels to progressively lower levels in the firm. Sooner or later in this process, allocations must be on a project, not a resource basis. It is the contention of this chapter that allocation to R & D is typically on a project and not a resource basis in the studies discussed here, owing to the fact that stable preferences for resources cannot be built up. Management seek to establish pattern in their perception of corporation–environment relations, and it is only when such pattern cannot be established that project allocations are made, sooner rather than later.

The open system interpretation also negates the criticism that 'rule of thumb' methods are 'illogical' or 'irrational'. Recognition and use of pattern need not imply articulation or analysis of possible reasons for the existence of a specific preference system; in fact it is difficult to see in what terms managers could justify or rationally explain what is a subjective non-analytic set of preferences based on experience. This especially the case if rational explanation is expected in terms of the neoclassical frame of reference.[8] Thus, Jewkes et al. (1969) find the non-articulation of reasons for budget determination in the modern corporation difficult to explain:

> One symptom of the difficulties of grafting research activities upon a business which must be guided by profit calculations is that firms seem not to know *how much* should be spent on research and their attempts to explain the grounds of their decisions usually seem to involve circular reasoning or to be inconsistent with known facts. . . . (p. 114).

The exasperation of Jewkes and his collaborators is directed with respect to a number of budgeting techniques, including 'percent of sales': 'Firms explain that they spend upon research some fixed proportion of their turnover without explaining how the proportion itself is determined' (p. 114).

However, 'explanation' in terms of the rational model is not possible in general. This need not pose severe problems since the conventional rational project-based model is redundant in constructing the abstract open system framework of corporate allocations.

SUMMARY

To summarise, the open system framework developed in the previous chapter may help provide a general explanation of differences between various budgetary methods employed in industrial allocations to research and development. The explanation of budgeting variation implies that large modern corporations utilising stable conventions or 'meta-rules' for allocating funds to respective functions are employing methods most appropriate to their circumstances.

It was also suggested that criticism of the apparent illogicality and circularity of such techniques was inappropriate when set in the context of the open system resource framework. The results of reinterpreting the status of various budgetary techniques in terms of the open system approach are therefore encouraging, and provide support for the resource base of such an approach. In the next chapter the framework will be extended to attempt to account for differences in research activity in various U.S. industries.

7 Distribution of R & D and Basic Research Activity in Industry

In this chapter we consider applications of the basic model with respect to R & D allocations in U.S. industry. The choice of U.S. industry is primarily a result of the availability of detailed information on the type and degree of R & D activity in U.S. manufacturing companies published annually by the National Science Foundation in the Surveys of Science Resources series. However, in the light of Chapter 6 we would also expect the applicability of the model to be greatest in U.S. industries; the importance and prevalence of large corporations in the United States provide greater opportunity for the development of stable resource preference systems than in any other economy.

For reasons discussed in a later section, the particular survey used is the 1963 survey (N.S.F., 1966). The sample used includes all manufacturing industries, and all non-manufacturing industries believed to conduct R & D. Sampling unit is the company (defined as all establishments under common ownership or control). Particularly important is that all companies with over 1000 employees were sampled with certainty. The few large companies which did not reply were sent a census mandatory form making clear the statutory obligation of firms to comply with the instructions of the survey, which was carried out through the offices of the Bureau of the Census.[1] Samples of firms employing fewer than 1000 employees were also taken. However, the data obtained on these small firms are patchy and irregular, and since our model is designed to be applicable to relatively large firms in any case, only firms employing more than 1000 employees are considered in this chapter.

In order to provide a framework for analysis of the N.S.F. surveys, in the next section the hierarchic approach of Chapter 5 will be adapted to the categories utilised by the N.S.F. Ideally, we would like data on basic research, applied research and development subsystems, but the N.S.F.

supply only data on basic research at disaggregated levels in different industries, and it is such a breakdown of subsystems data that is required to provide sufficient information for the subsequent analysis. Consequently, only a single boundary will be identified in the R & D system, that between basic research and the residual R & D subsystem. We shall also analyse conditions necessary for changes in industry distributions of R & D and basic research to take place, in order to provide a basis for the development and testing of hypotheses.

The approach used here is a cross-sectional one. Despite the fact that the basic formulation in the previous chapter emphasises adaptive learning to a particular steady-state value, time-series analysis is not used. A time-series analysis would tend to emphasise corporate *reaction* to changes in variables, which is not the main purpose of our study. Indeed, such change may be typically characterised as disequilibrium non-steady-state behaviour, in which a different set of decision-making procedures from normal steady-state assessment may come into operation. If the environment changes significantly and begins to pose problems for organisational survival, then the essentially reactive basis of time-series analysis may be emphasised by the triggering of problemistic search procedures and subsequent adoption of resulting solutions (such as crisis-provoked M-form organisation adoption in the 1930s). *The analysis of the determinants of a particular steady state need bear no obvious relationship to the analysis of the process by which that steady state is achieved* (as the concept of equifinality discussed earlier suggests). The analysis of the process of *attainment* of a new steady state, given disruption of previously viable steady-state solutions, is liable to involve a third-level negative feedback adjustment process leading to the eventual 'building up' of the new, viable, steady-state preference system by surviving firms. In such circumstances, received behavioural theory may provide useful concepts and techniques for analysis.

If, on the other hand, corporations have adapted to their requisite steady states, with the appropriate balance and distribution of funds to subsystems, a cross-sectional approach may be useful in analysing the determinants of steady-state behaviour. Provided we can specify how differences in intra-firm or environmental characteristics may affect steady-state allocations, we can use a cross-sectional approach to analyse differences in corporate allocations to R & D and subsystems in respective industries. This is the central concern of the next section, in which we discuss how changes in corporate environment may affect steady-state allocations.

CORPORATE REACTION TO CHANGES IN THE DETERMINANTS OF R & D STEADY STATES

The four-sectoral diagrams of Chapter 5 will be reduced to two sectors in analysing R & D and basic research together. We thus assume that management make only two decisions relating to budgetary allocations to technological change: firstly how much should be allocated to R & D rather than other uses of resources; secondly, how much should be allocated to basic research rather than applied research and development in the R & D system. The R & D budget thus constrains the choice of basic research and applied research and development allocations (see Fig. 7.1). The interpretations of basic research and R & D are those N.S.F. definitions referred to in Chapter 2 (see also Appendix I).

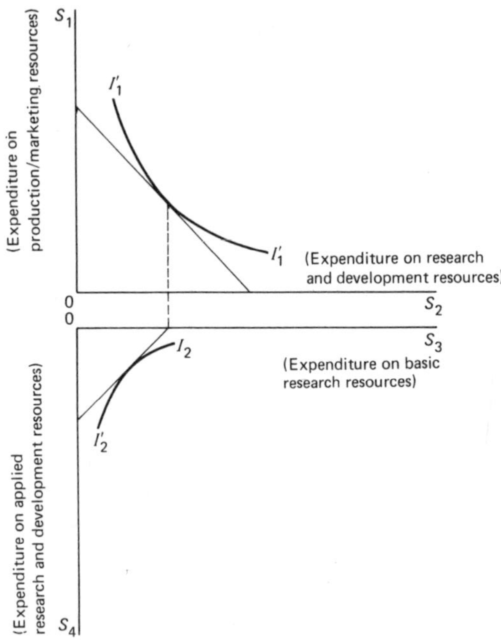

FIG. 7.1

The framework below can be easily utilised to explain differences in allocations between industries. From the concept of equifinality, and conditions for attainment of unique steady states in circumstances in which managerial discretion over resource allocation is not permitted, it follows that similar firms operating under similar conditions will allocate equivalent resource expenditures to each system/subsystem. It is worth re-emphasising that we assume that firms allocate resources *as if* survival is directly threatened; firms may possess what would otherwise be interpreted as organisational slack, but nevertheless maximise utility of resource allocation for fear of potential competition, take-over threats, etc.

In such circumstances, if a variable changes its value for some firms but not others, it may be that its effect will be to shift the preference system of affected firms relative to unaffected. In this way, differences in steady-state allocations may be observed. The problem may be formalised as follows. Suppose there is a differentiable variable α which operates on all firms but is an environmental parameter for each specific firm (environmental parameter is selected here for illustrative purposes, but the same argument holds for intra-firm parameters). In those circumstances the problem can be set up as one of constrained maximisation; in the first instance we shall be concerned with the S_1/S_2 decision:

$$\max U = U(S_1, S_2, \alpha) \quad (S_1, S_2 > 0) \tag{7.1}$$

$$\text{subject to } \gamma = S_1 + S_2 \tag{7.2}$$

where S_2 = expenditure on R & D resources; S_1 = expenditure on P/M resources; α = environmental parameter; γ = total available expenditure.

Using the technique of Lagrange multipliers:

$$U = U(S_1, S_2, \alpha) + \lambda(\gamma - S_1 - S_2) \tag{7.3}$$

where λ = Lagrange multiplier.

The first-order condition for a maximum requires:

$$U_1 - \lambda = 0 \tag{7.4}$$

$$U_2 - \lambda = 0 \tag{7.5}$$

$$\gamma - S_1 - S_2 = 0 \tag{7.6}$$

Taking total differentials:

$$U_{11}\mathrm{d}S_1 + U_{12}\mathrm{d}S_2 + U_{1\alpha}\mathrm{d}\alpha - \mathrm{d}\lambda = 0 \tag{7.7}$$

$$U_{21}\mathrm{d}S_1 + U_{22}\mathrm{d}S_2 + U_{2\alpha}\mathrm{d}\alpha - \mathrm{d}\lambda = 0 \tag{7.8}$$

$$0 - \mathrm{d}S_1 - \mathrm{d}S_2 = 0 \tag{7.9}$$

Dividing by dα and rearranging:

$$0 - \frac{\mathrm{d}S_1}{\mathrm{d}\alpha} - \frac{\mathrm{d}S_2}{\mathrm{d}\alpha} = 0 \tag{7.10}$$

$$\frac{-\mathrm{d}\lambda}{\mathrm{d}\alpha} + \frac{U_{11}\mathrm{d}S_1}{\mathrm{d}\alpha} + \frac{U_{12}\mathrm{d}S_2}{\mathrm{d}\alpha} = -U_{1\alpha} \tag{7.11}$$

$$\frac{-\mathrm{d}\lambda}{\mathrm{d}\alpha} + \frac{U_{21}\mathrm{d}S_1}{\mathrm{d}\alpha} + \frac{U_{22}\mathrm{d}S_2}{\mathrm{d}\alpha} = -U_{2\alpha} \tag{7.12}$$

Solving for $\mathrm{d}S_1/\mathrm{d}\alpha$ using Cramer's rule:

$$\frac{\mathrm{d}S_1}{\mathrm{d}\alpha} = \frac{\begin{vmatrix} 0 & 0 & 1 \\ 1 & -U_{1\alpha} & U_{12} \\ 1 & -U_{2\alpha} & U_{22} \end{vmatrix}}{\begin{vmatrix} 0 & 0 & 1 \\ 1 & U_{11} & U_{2} \\ 1 & U_{21} & U_{22} \end{vmatrix}} = \frac{\begin{vmatrix} 1 & -U_{1\alpha} \\ 1 & -U_{2\alpha} \end{vmatrix}}{\Delta} = \frac{U_{1\alpha} - U_{2\alpha}}{\Delta} \tag{7.13}$$

from the second-order conditions for maximum $\Delta > 0$. Therefore, the sign of $\mathrm{d}S_1/\mathrm{d}\alpha$ is the sign of $U_{1\alpha} - U_{2\alpha}$; similarly the sign of $\mathrm{d}S_2/\mathrm{d}\alpha$ is the sign of $U_{2\alpha} - U_{1\alpha}$.

The relative strength of cross-partial derivatives determines whether a steady-state allocation of R & D resources will increase or decrease as α varies. If α changes for a particular firm, then, if we have knowledge of the cross-partial derivatives, the direction of shift in resource allocation may be predicted, *ceteris paribus*. This may therefore serve as a framework for testing hypotheses based on expectations as to the value of $(U_{1\alpha}-U_{2\alpha})$.

This may also serve as a basis for analysing differences between corporate allocations to different functions. *Ceteris paribus*, variation in α between different groups of firms will result in different steady-state values of R & D and P/M, i.e.

$$(U_{1\alpha}-U_{2\alpha})\neq 0 \tag{7.14}$$

Similarly, the sign of $dS_3/d\alpha$ is the sign of $U_{3\alpha}-U_{4\alpha}$, and the sign of $dS_4/d\alpha$ is the sign of $U_{4\alpha}-U_{3\alpha}$. In this case, however, if eq. (7.14) holds, the budget constraint for S_3 and S_4 will vary also and must also be taken into account.

Figure 7.2 illustrates how differences in allocations may be analysed in terms of changes in parameters. Suppose the two indifference curves illustrated, $a'a''$ (firm A) and $b'b''$ (firm B), result from differences in α between the two firms. The firms otherwise operate under equivalent parameters and resource constraints. The two indifference curves may be interpreted as part of a continuous series along the resource constraint $r'r''$ created by variability in α between firms.

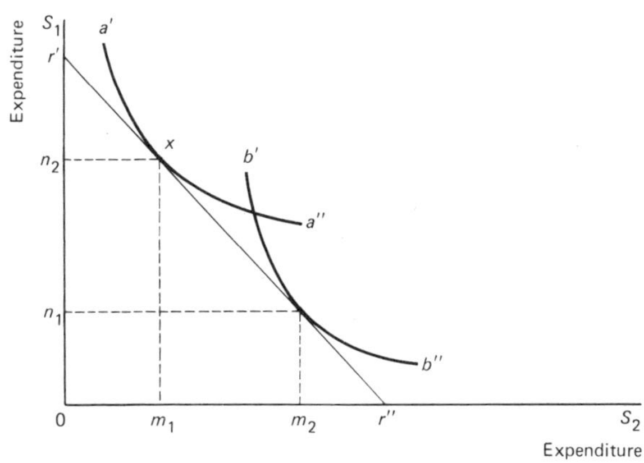

FIG. 7.2.

Firm A allocates On_2 resources to S_1 (P/M) and Om_1 resources to S_2 (R & D) at the point where utility of resource allocations is maximised subject to the resource constraint. Firm B operates under similar conditions to firm A (except for the α variable) and would allocate similar amounts to the R & D and P/M functions if α possessed the same value as for firm B. However, B operates under a higher value of α than A, and the relationship between utility and α is such that:

$$(U_{2\alpha} - U_{1\alpha}) > 0 \tag{7.15}$$

Consequently, the steady-state preference system of firm B is biased towards S_2 allocations relative to firm A's resource allocations; at point X the inequality in eq. (7.15) would encourage a redistribution in firm B's allocations until the appropriate steady-state allocation is reached at the utility maximisation point; in Fig. 7.2 this is where firm B maximises utility subject to $r'r''$, at On_1 of S_1 resources and Om_2 of S_2 resources.

This use of utility maximisation puts a different emphasis on sources of changes than does consumer theory; in the latter, tastes and preferences are generally assumed constant, and analysis is primarily concerned with changes in consumption caused by variation in prices and income. Here real prices and incomes are assumed constant, and changes in allocations due to changes in preferences are the prime concern of analysis. Consumer theory has tended to avoid this latter area of analysis;[2] as Havrilesky (1972) points out, economists have difficulty in talking about changes in tastes and their causes, 'by dint of trained incapacity' (p. 348).

The utility framework provides an extremely simple framework for the analysis of distribution of resources for R & D resulting from variation in the intra-firm environmental parameters. The direction of the predicted shift in steady-state allocations depends directly on the hypothesised sign of $(U_{1\alpha} - U_{2\alpha})$. In the next section we shall consider hypothesised effects of a number of variables using this mode of analysis.

DETERMINATION OF R & D STEADY STATES: HYPOTHESES

A recurrent feature of studies in the institutional and behavioural literature on technological change is that although formally such studies may give lip-service to neoclassical theory or its derivatives, informally investigation of behaviour often proceeds in terms of a resource-based interpretation of corporate behaviour. Not surprisingly, such work often sits uncomfortably in its neoclassical frame, and is vulnerable to criticism couched in neoclassical terms. However, the fact that such informal and descriptive interpretations are made in apparently neoclassically based works facilitated the development of the basic model in an earlier chapter. In this context, it should not be surprising if the literature supplies possible hypotheses interpretable in resource-based terms; indeed, it would be surprising if it did not. We shall consider a number of different variables in turn in terms of their likely or expected effects on distribution of R & D activity in the firm. In the first instance we shall restrict consideration to R & D activity at the gross level of P/M and R & D systems and, where appropriate, compare the resources hypothesis with the corresponding neoclassical hypothesis.

At this point it may be appropriate to clarify the relationship between the systems framework and the hypotheses of this and later sections. As was emphasised in Chapter 1, we do not attempt to test or refute directly the open system framework; instead its possible relevance and usefulness will be assessed by the performance of the lower-level hypotheses based on this systems approach. We distinguish between higher- and lower-level hypotheses in that higher-level hypotheses are used to explain lower-level ones, the latter being derived from the arguments of higher-level hypotheses. Typically, the higher the level of a hypothesis, the less the chance that it contains terms relating directly to observables (see Cyert and March, 1963, pp. 299–300). Thus, in our analysis, the highest-level hypotheses such as utility maximisation and the hierarchical model of the previous section do not contain empirically observable terms. Instead we must obtain lower-level hypotheses derived from, and explained in terms of, the arguments and concepts of our resource utility framework.

In this respect our interest is in how the regression analysis performs *as a whole* in attempting to explain differences in R & D activity. Failure in one part of the analysis need not imply automatic revision or rejection of the systems approach. In the following analysis some hypotheses tested find no support from the regression analysis, at least in the form in which they are operationalised here. Our concern is a more general one in that we are concerned with developing an approach within which we can formulate useful hypotheses; within this framework some hypotheses may be more obvious or strongly held than other more speculative or tentative ones.

In this analysis the status of a variable and its expected effect on R & D activity compared to, say, neoclassical analysis may differ as a consequence of differences in interpretation of the frames of reference. Also the significance or otherwise of a variable in the regression analysis may be strongly affected by the presence or absence of other variables in the regression equations. In such cases, statistical significance in the relationship between particular independent and dependent variables may only result after separating out the effect of other independent variables. For both these reasons, while previous empirical analysis has relevance to the present one, comparability between previous studies and the present one must be limited to the extent that these factors have importance. Consequently, while a review of past empirical analysis is essential to put the analysis of this chapter in context, previous empirical studies have a limited contribution to the present analysis (with exceptions specified below) and therefore are mainly contained in an appendix to this chapter.

(a) *Technological opportunity*

Nelson et al. (1967, p. 73) argue that industries differ substantially in their capability for invention and suggest that high R & D to sales ratios are partly a consequence of a greater ease of achieving technological advances in those industries, attributing this to the science base of those high 'technological opportunity' industries. Brozen (1965, pp. 92–9) and Mansfield (1968a, p. 59) make similar assertions, and Brozen statistically demonstrates the higher R & D intensity in science-based industries using N.S.F. data.

Using different measures or indices of technological opportunity or progressiveness, Phillips (1966), Comanor (1967) and Scherer (1967) all found a strong positive relationship between their particular measure of technological opportunity and research intensity. As far as interpretation and comparison of these studies is concerned, it must be emphasised that measures or estimates of technological opportunity are difficult to derive, since qualitative differences may complement quantitative differences in opportunity; one industry may utilise a scientific and technological base which intermittently throws up a radically different and significant invention, while another could annually produce a regular yield of numerous minor patents. Comparability between technologies and industries is difficult to achieve in many cases; we shall therefore restrict our interpretation of technological opportunity to that of Phillips (1966), who defined technological opportunity (P_j) as a 'subjective evaluation of the extent to which current science permits functional (as contrasted with stylistic) product changes and product differentiation among firms' (p. 305). Phillips' variable will in any case be utilised in later regression analysis of distribution of R & D activity.

A simple hypothesis is suggested by looking at P_j in a resources context. If P_j is interpreted as α, then the effect of P_j on resource allocation may be analysed by reference to the earlier utility-maximising model. In this model, dS_1/d_α is the sign of $(U_{1\alpha} - U_{2\alpha})$;[3] we therefore have to compare the possible signs of the partial derivatives of $\partial U/\partial S_1$ and $\partial U/\partial S_2$, both with respect to α.

It is reasonable to assume that $U_{1\alpha}$ is zero. We would not expect a technological change parameter directly to affect resource utility in production and marketing; instead, we would expect the direct effect of P_j to be localised on resources for technological change, R & D. As far as $U_{2\alpha}$ is concerned, we would expect the cross-partial derivative to be positive; increasing P_j would tend to increase $\partial U/\partial S_2$: as technological opportunity increases, so does the marginal utility of resources allocated to R & D, since the increased ability of R & D resources to create functional[4] product changes enhances the contribution R & D resources make towards achievement/maintenance of steady state. Therefore:

$$(U_{1\alpha} - U_{2\alpha}) < 0 \therefore \frac{dS_1}{d_\alpha} < 0 \quad \text{and} \quad \frac{dS_2}{d_\alpha} > 0$$

The net sign of dS_1/d_α is negative, similarly that of dS_2/d_α is positive.

The effect on corporate preference systems of P_j is summarised in Fig. 7.3. The direction of the arrows indicates the shift in corporate allocations as P_j increases, given $r'r''$. At r' the technological opportunity is non-existent or so low as to not encourage any allocations to R & D; all allocations are to production/marketing. As P_j increases, resources are allocated to R & D and the amount progressively and proportionately increases as P_j increases.[5] We might reasonably expect a limit to the proportion of R & D resources since some P/M resources will be required to exploit the R & D output; this is designated by the nominal combination *Y* in Fig. 7.3.

Technological opportunity may be interpreted as an environmental parameter operating on firms in a particular industry. Its effect in resource utility is learnt by feedback from past resource allocations. It may thus be utilised to explain differences in resource allocations in different industries, since typically different industries have differing P_j values associated with them.

However, the concept has certain difficulties when applied in a strict neoclassical framework in which optimal combinations of resources may be determined. Such a framework appears implicit in the analysis of Nelson et al. (1967):

> One would not expect the same ratio of R & D to sales to be optimal in all industries and product fields. . . . In the first place, returns on

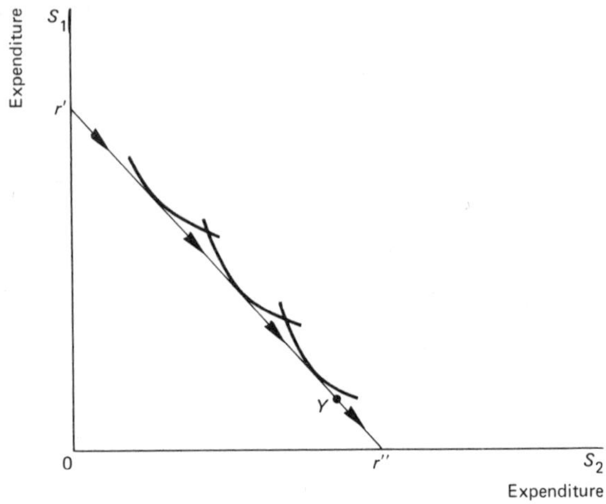

FIG. 7.3.

> R & D aimed at product improvement or new products depend not only on the level of sales, but also on the value consumers place on improved product performance. . . . In the second place . . . industries differ significantly in their capability for invention. (p.73)

However, as Minasian earlier stated (1962):

> 'Opportunity' is [a] hypothesis which states that research and development expenditures are determined by . . . the technical possibilities which may vary among firms. Therefore, the firm will indulge in such activity only if its production processes readily lend themselves to improvement. This seems illogical. If returns are high to research and development for particular product mixes, a firm should (and would) carry on such activity whether it produces such mixes or not, as it can alter its mix to take advantage of the resulting improvement, or sell that improvement to a firm that can use it directly. (p. 122)

In the neoclassical frame of reference, Minasian is perfectly correct. In general, under conditions of perfect knowledge and perfect mobility of resources, there are few barriers to R & D being conducted in the firm;

even if barriers to entry do exist in implementing the derived invention, it should be possible to sell or license the invention to other firms. The assertion of Nelson et al. (1967) above is valid if R & D activity is restricted to certain fields (industry or product), but in general there is no reason why this should be the case, since the direction of R & D can easily be reoriented to new areas of promise at low cost. As has already been pointed out, Nelson et al. themselves identify the high degree of R & D diversification typifying most firms throughout U.S. industry (see pp. 50–2).

It may help if we use the resources model to examine this problem. The resources model was partly based on the idea of separation of budget and direction of allocations within the budget; allocations to particular functions were determined by managerial learning of the characteristics of resource allocations from past allocations. While the direction of new projects will tend to depend on hunches and perceived opportunities, the resource constraint itself depends on *retrospective* evaluation of system performance, and consequently tends to be determined by history and existing system and environmental parameters.

Yet this does not explain how a steady state may exist in low P_j industries. What is to stop one firm racing ahead of its competitors through becoming research-intensive and adopting a highly diversified research programme? The concept of synergy adopted by Ansoff (1965a) may provide some guidance in this respect. According to Ansoff, synergy is the 'effect which can produce a combined return on the firm's resources greater than the sum of its parts . . . frequently referred to as "2 + 2 = 5"[6]. A number of types of joint effects or synergy may operate on anticipated combinations of resources and projects, such as common distribution channels, joint use of equipment, etc., but one of the most important types is management synergy. Management face different sets of problems in different industries and if the new areas produce similar problems to those encountered in the past, management may usefully integrate the new activities into the enterprise (Ansoff, 1965a p. 76).

However, the opposite effect may prevail: managerial competence, experience and attitudes may inhibit or interfere with adoption of new activities if it attempts to transfer or apply management techniques learnt in one context to a new area where it is inappropriate. Negative synergy may exist, in that the resource characteristics of corporations may inhibit it from entering new promising areas; as Ansoff points out (p. 74), a firm in the defence industry would be at a disadvantage if it attempts, without prior experience, to enter into a highly competitive consumer area such as the tobacco or motor industries. Firms may be 'allergic' (Loasby, 1967, p. 301) to application of resources in highly novel areas.

Consequently, for a particular resource profile of a firm operating in a particular environment, there are liable to be negative as well as positive synergic relations with other environments. As with the Penrosian model, existing resources both enhance and inhibit different areas of R & D activity. Therefore a steady state for corporations operating in one environment may be substantially different from that for corporations operating in another environment.

We can summarise the main arguments made here between the neoclassical and resource-based approaches as follows. In a neoclassical context, as explained by Minasian above, the central concept is that of the product or project. Assuming firms are efficient and have perfect knowledge, they will adopt an opportunity which is expected to generate profits. If firms differ in their eagerness or ability to take advantage of a particular opportunity, the obvious inference is that they differ in efficiency terms; this is a natural consequence of assuming that product mixes, production functions and projects may be considered separately from the resource profile of the firm.

However, in our resource-based approach, technological opportunity may create differences in steady-state allocations for rational decision-making. Firstly, it creates different experience of past utility of R & D resource allocations in different environments through feedback effects, and in consequence results in differing values placed on R & D allocations according to the context. Secondly, in this approach, the existence of the resource may precede that of the project. This serves to emphasise that assimilation of the opportunity may depend crucially on its relationship to existing products and resources in the firm.

Thus, while the highly specialised nature of the resources and operating experience of a particular corporation or industry sector may be an inhibiting factor restricting realisation of opportunities in other sectors, it may also act as a barrier the other way in preventing outsiders from exploiting opportunities in that sector. It is worth emphasising, however, that if a relevant breakthrough, or series of breakthroughs, is made elsewhere, the sector's natural resource advantage may be an insufficient barrier and it may be vulnerable to attack from outside (such as chemicals invasion of textiles in the post-war period).

In short, in the resource-based approach, differences in research intensity may be explained as a *rational* consequence of differences in the resources and technological opportunity of respective sectors. Neoclassical theory is deficient in this respect because of the product/project emphasis of its analysis.

(b) *Growth*

Again, the neoclassical interpretation of the effect of this variable in R & D projects is vulnerable to criticism within the same frame of reference. Mueller (1967) suggests: 'The faster a firm's sales are increasing, the more confidence it will have about its ability to secure the benefits from uncertain R & D projects, and the more patience it can afford to show in waiting for these benefits. The faster a firm's sales are growing, the greater economic advantage it receives from a given cost-reducing invention' (p. 73).

However, by the same argument, the faster a firm's sales are growing, the greater economic advantage it may also expect from simple expansion of existing product lines; how is it possible to derive unambiguous predictions of variation in steady states as growth of firm or industry changes?

At the higher level of abstraction associated with the resources model, the problem may be reset in a resources context. From the point of view of the individual firm, it is reasonable to suppose that innovations will be more easily accepted and have a greater chance of success if they can be aimed at new market areas rather than old; habit, inertia and apathy can make it difficult for new products to replace old ones, while new processes may find it difficult, if not impossible, to replace old processes if these are already installed. The section of demand in which innovations enjoy comparative advantage is the *new* demand in which both innovations and existing products compete on equal terms. Expected high growth of the industry would imply favourable conditions for the exploitation of innovations, permitting innovation 'Lebensraum' and associated increased utility of R & D resources. In the limit, if no new demand is being created at all, innovations will have to compete with existing products already advantageously 'occupying' the demand space. On the other hand, growth would create new market areas in which no existing products would have the natural advantage of market 'possession'.

If we interpret recent growth of the industry as being the relevant environmental parameter (α) affecting perceived utility of resource allocations, then we may interpret the above argument as signifying

$$U_{2\alpha} > U_{1\alpha}$$

That is, as growth of industry increases, $\delta U/\delta S_2 > \delta U/\delta S_1$: growth tends to have a greater effect on the marginal utility of R & D resources than on that of P/M resources. Consequently, $dS_1/d\alpha$ is negative and $dS_2/d\alpha$ is positive.[7]

A study by Freeman (1962) may be cited as supporting the industry growth hypothesis; industrial research expenditures as a percentage of net output for 17 U.K. and U.S. industries in 1958 were found to be highly correlated with the past growth of output of respective industries from 1935 to 1958. The correlation coefficients were 0.95 for the U.K. industries, and 0.74 (product basis) and 0.76 (company basis) for the U.S. firms. While Freeman is cautious as to the causal significance of this association, the fact that *past* growth is the variable associated with industrial R & D expenditure would tend to suggest that it is growth facilitating R & D, if there is a direct relationship between them.

However, already the interrelationships between variables may turn out to be complex and confused. In a high P_j industry we would expect research intensity, which in turn we could expect would lead to industry growth; if P_j is an enduring variable, the industry would have high P_j, high growth and a high level of research expenditures. While the problem of direction of causality may be eased by considering association between past hypothesised independent variables and present research expenditures, separating the effects of just the two variables already considered would appear endangered by the likelihood of multicollinearity. This problem notwithstanding, we would expect growth to lead to high levels of research expenditure, *ceteris paribus*.

(c) *Federal Funds for Research and Development*

The preceding two variables are interpretable as *environmental* parameters, technological opportunity and growth of a particular industry acting on the corporate preference system from the outside environment. However, the steady state is determined by the relationship of environmental and intra-firm parameters, and since we shall be concerned with the allocation of resources in U.S. industry in 1963, one potentially important intra-firm consideration is federal financing of R & D.

Rankin (1956) outlines the system of federal financing of R & D. The authorisation of government agencies, and frequently the area of research itself, is specified by Congress. Research proposals not able to be dealt with by a particular agency may be contracted out to private industry. Black (1969) points out that federal funds are usually allocated for clearly defined projects and that 'level of effort' work is supported infrequently (pp. 216–17). Collier (1963) asserts that R & D workers hired for federal contracts tend to be laid off when business falls off, and hired again when it picks up with the signing of another big contract.

Our prime concern is with *corporate* decision-making and allocations to R & D, and it is in terms of company-financed R & D that the model and subsequent discussion was based. If Collier's interpretation of the behaviour of firms in taking on federal contracts is true, then we might expect to find that the functional allocation of corporate-financed R & D resources is little affected by federal projects taken up by the firm; resources are employed in federal projects when funds are available and not, when funds dry up. This suggests minimal substitutability of federal and company resources; therefore:

$$U_{1\alpha} = U_{2\alpha} = 0$$

where α is percentage of total company R & D federally financed.

However, there exist opposing arguments:

> A counterbalance to any advantage to industry from government research is the very real danger that government research will weaken the effective research supported by private enterprise upon which our industrial strength has been based. (Boundy and Chamberlain, 1959, p. 83)

> The weapons industry is doing less and less company sponsored research and relying more heavily than before upon federal agencies to provide the initial guidance for development efforts. (Horne, 1962, p. 329)

> Industry is being priced out of the [R & D] market by the competition of government. (Brown, 1962, p. 359)

> There is a danger that the government may be becoming an overwhelming competitor in some areas of research. It may expand its activities in particular fields to such an extent that the field is no longer attractive to private research. The result might be massive research and development expenditures with little resulting commercial developments . . . increasing government activity in specific fields does not usually result in increased *total* activity in those fields. It may result merely in a reduced amount of *private* activity. (Ellis, 1962, p. 369)

Further, Brozen (1962, p. 275) mentions that evidence on federal contracting in the 1950s suggested that government R & D financing simply replaces private R & D resources, Orlans (1973) suggests that government funds for space and weapons R & D retards the growth of privately financed R & D, and Black (1969) in this context speculates that private firms search out federal agencies with research objectives comparable to their own to finance R & D that the company might otherwise have had to finance itself.

These all suggest that federal allocations might substitute corporate resources for R & D, i.e.

$$U_{2\alpha} = -\text{ve}$$

In such circumstances:

$$U_{1\alpha} - U_{2\alpha} > 0 \;\therefore\; \frac{dS_1}{d\alpha} > 0$$

$$U_{2\alpha} - U_{1\alpha} < 0 \;\therefore\; \frac{dS_2}{d\alpha} < 0$$

As federal financing of R & D increases, the company would tend to employ fewer corporately financed R & D resources. In the above we assume that the resource constraint does not vary, and the federally financed R & D resources are classified along with the residual P/M resources.

We shall therefore include a variable measuring federal financing of R & D at industry level to examine which of the alternative hypotheses fits the observed behaviour of industrial allocations better.

(d) *Q.S.E.s*[8] *as Proportion of R & D Expenditure*

Frequent mention is made in the R & D literature of the central importance of the human element in R & D. According to Carter and Williams (1959):

> in organising research it should be remembered that the individual scientists are generally the 'scarce resources' round whom, and in the light of whose needs, the organisation should be built. Hence the common insistence that 'men are more valuable than equipment'. (p. 50)

Also Rath (1967):

> People, facilities and 'knowledge' are key resources of all research and development communities. Of the three, people are the most significant and critical resource. People are necessary to design, construct, modify and operate facilities. People are the main instrument for production, transmission and retrieval of 'knowledge'. (p. 104)

This strongly suggests that the qualified scientific and engineering labour component in R & D is of potentially higher utility than the capital element. While formulation of such assertions in an acceptable operational form is liable to be extremely difficult, omission of consideration of possible asymmetry in utility derivable from R & D resources may lead to an important determinant of R & D expenditure being overlooked. Technological factors are liable to determine the relationship between employment of scientists and engineers, reducing the discretion management have over the capital/labour composition of the R & D resources 'bundle'. The inference of the statements by Carter and Williams, and Rath, is that for any two firms having equivalent 'bundles' of R & D resources so far as total expenditure is concerned, the firm with the greater human element in its R & D budget will be expected to derive the greater utility from its R & D allocations. Obviously the importance of the 'human element' might be measured in different ways; one way is to assume that the significance of qualified R & D scientists and engineers relative to other R & D resources can be measured by the wages and salaries of scientists and engineers as a percentage of total R & D expenditure. The statements above suggest:

$$U_{1\alpha} = 0, \; U_{2\alpha} > 0$$

$$\frac{dS_1}{d\alpha} < 0 \text{ and } \frac{dS_2}{d\alpha} > 0$$

where α is $\dfrac{\text{Wages and salaries of Q.S.E.s}}{\text{Total R \& D expenditure}}$

This hypothesis must be interpreted as being particularly speculative and tentative, especially since it is an intra-firm variable measured at industry level owing to the limited data availability.

(e) *Size of Firm*

Size of firm is a variable which differs significantly from those preceding; whereas the other variables may be interpreted in terms of the variable α, size of firm relates directly to the resource constraint $r'r''$. What we are concerned with here is the expansion path of the firm in terms of resource allocation, and the effect that available resources have on the distribution of resources within the firm. However, our approach will be similar to the extent that we shall be looking for asymmetry of effect on resource utility. If variation in available resources has differing implications for R & D and P/M functions, it may be possible to frame testable hypotheses as to possible relations between size and research intensity.

In fact, following Schumpeter (1954), a number of studies have attempted to establish a connection between size and research intensity, but results have not been consistent,[9] and to date no firm conclusions have been obtainable. This need not imply that revision of Schumpeter's theory of 'creative destruction' is necessary, since Schumpeter was primarily concerned with innovations, the final output of R & D departments, not the process of producing those innovations.

Given these results of research on effect of size of firm on R & D expenditures, what justification is there for further analysing its possible effects in this context? Apart from the reason that omission of consideration of size of firm might distort results if it does turn out to be a significant factor in determining allocations, the primary reason is that inclusion of the other variables in the analysis may separate out effects obscuring the true relationship between research intensity and size. However, there are also sound reasons why we might expect research intensity to increase with size of firm as a result of differing conditions for innovation.

The best conditions for innovation are often found in small companies where communications between development, production and marketing are simple and easy, and where a common objective can be easily established (Layton, 1972, p. 5). In such circumstances, the small firm may survive even in turbulent environments, by relying on rapid adoption or imitation of new techniques. However, the large firm usually has more complex and potentially conflicting issues and problems to deal with. A major change of direction for a large firm may require processing information and decisions through a large number of levels in the firm hierarchy and co-ordinating a large number of subsystems in the change process. Whereas intra-firm mobility of resources is relatively easy to achieve for a small firm, it may be a severe problem for a large firm.

Consequently the small firm may have an advantage at the innovation end of the spectrum of management of technological change. However, as the firm increases in size, the difficulties of co-ordination and reorientation of resources with respect to *adaptation* to external change will tend to increase the attention paid to *anticipating* environmental change – organising R & D, technological forecasting, strategic planning, etc. Further, not only is the large firm more liable to rely on R & D as a necessary instrument for technological change, it is also more able in that, as was suggested in the preceding chapter, it may establish more easily a resource-based overview of the R & D function. Individual projects do not dominate the strategic consideration of the enterprise, nor does each project entail committing a high proportion of organisational resources in highly uncertain activity, unlike in small firms.

Thus, the small firm may achieve a steady state by acting as a *reactive* system rather than by organising resources in an R & D subsystem and itself instigating major technological innovation; however, the large firm, because of its relative disadvantage in changing direction, is more liable to attempt to anticipate environmental change by incorporating an R & D department in its operations. This would tend to suggest that typically there will be a switch in emphasis from a dependent, reactive strategy to a longer-term R & D based innovative strategy, as the corporation grows in size. This interpretation is supported by the observation that in 1953, in the United States, almost all firms employing more than 5000 workers undertook R & D work, but only about 10 per cent of firms employing fewer than 5000 workers did, with a steady gradation between these extremes (Jewkes et al., 1969, pp. 123, 147).

This argument therefore suggests that research intensity will tend to increase with size of firm. Thus, as a firm increases in size (measured in terms of available resource expenditure), the proportion of resources allocated to R & D will tend to increase. This is not to suggest that large firms in a particular industry are more innovative than small firms; the similar *I* variables are liable to force innovativeness on all firms operating in hostile environments. Rather it implies a switch in emphasis from innovation *adoption* to innovation *generation*.

This concludes the set of variables to be considered as possible determinants of R & D activity. In a later section we shall consider how they may be operationalised in the regression analysis; for the moment we shall go on to consider similar possible relationships as far as basic research is concerned, with the reminder that the hypotheses developed in the context of the systems approach in this section are possibly only a subset of those conceivable under the systems framework. The justification for those selected remains their apparent reasonableness within the systems frame of reference compared to other conceivable relationships considered, but rejected, for the purposes of the regression analysis.

DETERMINATION OF BASIC RESEARCH STEADY STATES: HYPOTHESES

As far as the basic research subsystem is concerned, the allocation problem is similar to that of R & D and P/M distribution, except that in this case it is the size of R & D establishment which constrains the distribution of resources to the subsystems, not size of the firm itself (see Fig. 7.4). The dotted line at the top of Fig. 7.4 indicates the transmission of available resources to R & D (interpretable as size of research establishment) to provide the constraint for the allocation problem at the

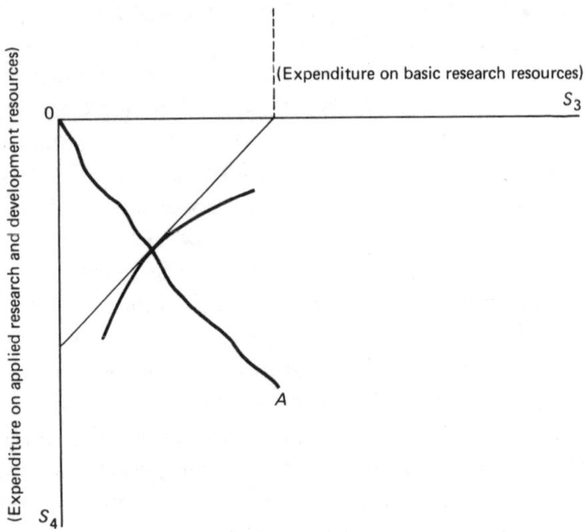

FIG. 7.4.

next level down. What was an endogenous variable for the R & D–P/M allocation problem is an exogenous parameter for the basic research/applied research and development problem.

The behaviour of the variables in this problem may also be similarly treated as in the higher-level R & D–P/M problem. The decision is:

$$\max U = U(S_3, S_4, \alpha); \quad S_3, S_4 > 0$$

$$\text{subject to } Y = S_3 + S_4$$

where Y = size of research establishment measured in total available expenditure on resources.

In an exactly similar fashion to the S_1, S_2 allocation problem:

$$\frac{dS_3}{d\alpha} > 0 \quad \text{if } U_{3\alpha} - U_{4\alpha} > 0$$

and

$$\frac{dS_4}{d\alpha} > 0 \quad \text{if } U_{4\alpha} - U_{3\alpha} > 0$$

(a) *Technological opportunity*

Phillips' index P_j provides values for each industry 'purporting to measure the strength of the association between organised sciences and the technologies of the respective industries' (Phillips, 1966, p. 305). A reasonable hypothesis in such circumstances is that $U_{3\alpha} > U_{4\alpha}$, where α is P_j of the appropriate industry; since an 'applied research'·linkage is necessary for 'basic research' to be processed through to final output, high P_j would imply relatively easy transfer of basic research. Low association between science and technology would imply poor linkage between basic and applied research. In the limit when P_j is zero, so also would be allocation of resources to basic research. Therefore we may reasonably expect P_j to have a proportionately greater effect on the resource utility in the basic research subsystem than the residual R & D subsystem; therefore $dS_3/d\alpha > 0$ and $dS_4/d\alpha < 0$.

(b) *Growth*

It was argued earlier that industrial growth favoured R & D activity since new products and processes tended to be disadvantaged in competition with old products for existing markets. This argument holds even more strongly with respect to the typical final output derived from allocation of resources to basic research; the output derived from basic research tends to be highly novel and unfamiliar, to an even greater extent than R & D output resulting from the residual R & D subsystems. We would therefore expect that resistance to the final output derived from basic research would be greater than to final output from subsystems nearer the 'bottom end' of the R & D spectrum. In such circumstances, basic research would be even more dependent on industrial growth than R & D work in general; *ceteris paribus*, it would be more difficult for innovations derived from basic research to make inroads into stable, existing markets than it would be for less radical output, and consequently a specific level of industrial growth should have a stronger effect on the expected utility of allocations to basic research than on the expected utility of allocations to other R & D subsystems. Consequently, we would expect $U_{3\alpha} > U_{4\alpha}$, and $dS_3/d\alpha > 0$, $dS_4/d\alpha < 0$

(c) *Federal Funds for Research and Development*

Black (1969) earlier pointed out that federal funds for R & D tended to be allocated for specific projects; we would consequently expect that as the proportion of federal funds for R & D rises, basic research projects (N.S.F. definition) would tend to be squeezed out of the R & D subsystem, interpreting both basic research and the R & D system itself in terms of total resources (company and federal).[10] This suggests the simple hypothesis that $dS_3/d\alpha < 0$ and $dS_4/d\alpha > 0$ (where α is percentage of total company R & D federally financed) owing to the application-specific bias of federal funding.

(d) *Q.S.E.s as Proportion of R & D Expenditure*

As with the P_j and growth variables, the arguments put forward as to the possible directional effects of labour intensity of R & D on the resource utility of that function hold even more strongly for basic research. As Mansfield (1968a, p. 47) suggests, the development end of the R & D spectrum is typically more expensive than research; building of pilot projects, construction of new materials and prototypes can be extremely costly, and tend to imply relatively high capital costs relative to labour. Basic research activity on the other hand does not generally require such expensive capital but instead relies on quality of scientists and engineers. Interpreting α as proportion of R & D costs spent on wages and salaries of scientists and engineers, this would tend to suggest:

$$U_{3\alpha} > U_{4\alpha}$$

$$\therefore \frac{dS_3}{d\alpha} > 0 \text{ and } \frac{dS_4}{d\alpha} < 0$$

(e) *Size of Research Establishment*

This variable has a rather different status from the others discussed since it is concerned with how research intensity may vary with size of subsystem (*ceteris paribus*). Figure 7.4 illustrates an example in which the proportion of R & D resources allocated to applied research and development increases along the expansion path OA, indicating that intensity of basic research activity declines with size of research establishment.

In fact we would expect the basic research share of R & D to *increase* with size of research establishment. It is likely that the utility of basic research allocations will increase more rapidly than those of applied research and development as size of R & D department increases because of increased range and variety of skills and resources afforded for the exploitation of basic research. As was emphasised in Chapter 2, the links between science and technology are often weak and indirect, and the embodiment of basic research in a specific invention is not envisaged at its onset. Consequently a small R & D department may have only limited skills, abilities and training to exploit the innovative potential resulting from a specific in-house basic research project. On the other hand, the possibilities suggested by the same project would be more likely to be successfully exploited by a larger R & D department with a wider range of resources and techniques. Since the specific commercial opportunities resulting from basic research projects are unexpected, the larger R & D department has a better chance of putting together the team necessary to exploit it satisfactorily.

This has similarities to Nelson's (1959a) widely quoted diversification hypothesis: 'a firm producing a wide range of products resting on a broad technological base may well find it profitable to support research toward the basic-science end of the spectrum . . . a broad technological base insures that, whatever direction the path of research may take, the results are likely to be of value to the sponsoring firm' (pp. 302–3). Like Nelson's hypothesis, the above emphasises the ability of the corporation to take advantage of the unexpected avenues opened up by basic research, though our hypothesis relates to the capabilities and range of R & D resources, Nelson's hypothesis to the capabilities and range of production resources. Essentially the only difference between the two hypotheses is the stage of the innovative process on which emphasis is placed.

EMPIRICAL EVIDENCE OF STEADY STATE IN INDUSTRIAL R & D ACTIVITY IN THE UNITED STATES

To test these hypotheses, the 1963 survey of R & D in U.S. industry (N.S.F., 1966) is utilised. As mentioned earlier, cross-sectional analysis is used since we assume that firms are operating in a steady state and have achieved a match between themselves and their environment.[11] We shall examine whether differences in industrial R & D intensity are consistent with the hypotheses above.

The year 1963 was chosen primarily because there is evidence to suggest that R & D activity in the United States stabilised around that period, in terms of allocations relative to alternative uses of resources. Figure 7.5 illustrates the growth of R & D activity relative to G.N.P. at national and industrial level from 1953 to 1970.

The development of industrial R & D as a percentage of G.N.P. continues, and partly overlaps, the trend outlined by Clare in Fig. 6.1 (see p. 121). The overall trend suggested by the two graphs considered together is of an S-shaped curve development which levels off at a plateau observable from 1960 to 1964 (Fig. 7.5). We are interested in the possibility that the economy overall demonstrates steady-state behaviour in allocations to R & D as opposed to other functions, i.e. that a 'balance of resources' is obtained with a constant ratio between R & D and G.N.P. In this case it would be in the early 1960s when the sigmoidal growth patterns appear to reach saturation level. There is some tailing-off of R & D performance after 1964, and we deal with this point below.

As far as individual industries are concerned, there is less consistency in the pattern of R & D funding. A useful summary is again provided by the N.S.F. for the 1958–71 period in terms of the number of R & D scientists and engineers per 1000 employees in five leading industries (Fig. 7.6). The period up to 1961 is typified by erratic growth in the proportion of R & D scientists and engineers in each industry; the period after 1965, by decline in the proportion employed by the three industries of the group employing the highest proportion of R & D scientists and engineers. However, the period 1961–5 appears to exhibit stability in the

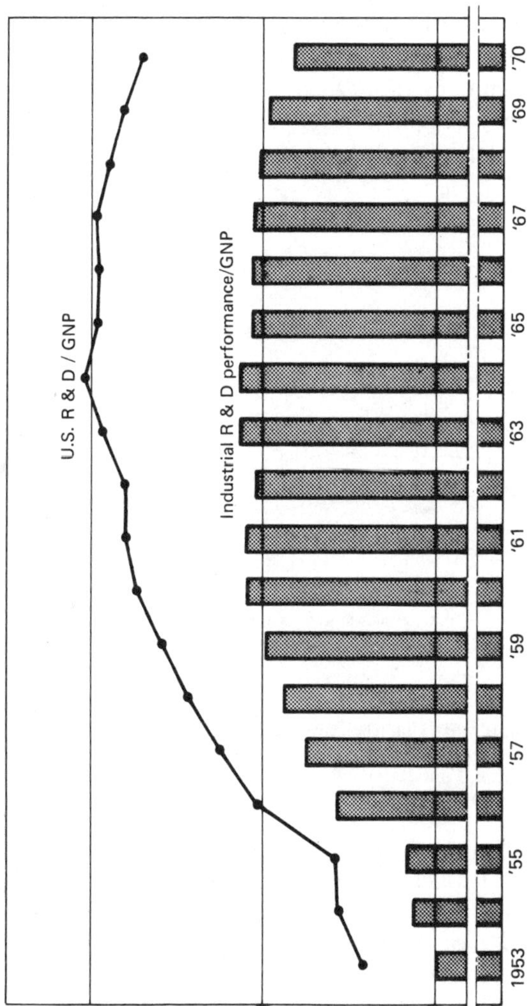

FIG. 7.5. R & D as proportion of G.N.P., 1953–70.

Source: National Science Foundation, *Research and Development in Industry 1970*, Surveys of Science Resources Series, Washington, 1982, p. 3

proportion of employees that are R & D scientists and engineers in each industry, with the exception of aircraft and missile in which the proportion continued to grow strongly until 1965, after which it declined sharply. This change of trend may be attributed to cutbacks in the NASA space programme and federal defence funding. With the repercussions felt in R & D work in closely linked industries, such as electrical equipment and communication, these major switches of emphasis in R & D funding may be indentified as being primarily responsible for both the sectoral instability and the change in trend of total R & D funding after 1965.

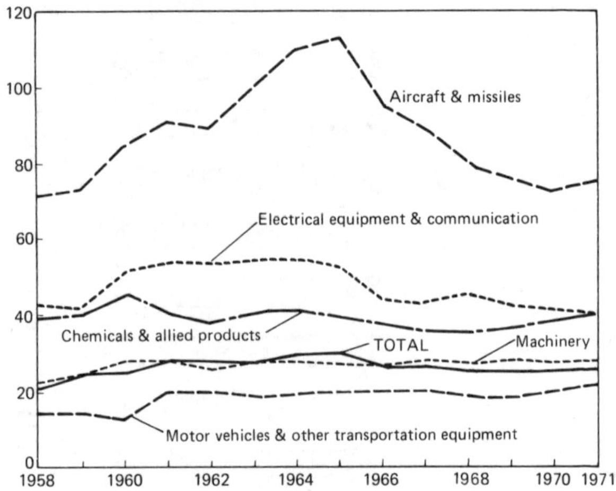

FIG. 7.6. Number of R & D scientists and engineers (full-time equivalents) per 1000 employees in five leading industries, 1958–71.

Source: National Science Foundation, *Research and Development in Industry 1971*, Surveys of Science Resources Series, Washington, 1973, p. 7.

Consequently the 1961–5 period, and in particular the 1963–4 period, appears the point at which the U.S. economy as a whole most closely approximated steady-state behaviour with respect to, and in comparison with, resource allocation in other activities. During this period, performance of industrial R & D as a percentage of G.N.P. appears to reach a plateau level after its sigmoidal growth pattern, while subsequent gross and sectoral funding of industrial R & D appears to be substantially affected by changes in federal policy with respect to funding of industrial R & D – implying non-steady-state behaviour. Consequently, 1963 was chosen as a suitable year for the analysis, a further point in support of this choice being that growth indices are available for the 1958–63 period,[12] in terms of change in production of each S.I.C. industry. It is the value of independent variables (such as growth) in the year *preceding* the measurement of the dependent variable involving R & D activity that we wish to utilise, and so 1963 is a particularly suitable year for this reason also.

MEASUREMENT OF VARIABLES USED TO TEST HYPOTHESES

The hypotheses developed earlier are typically of a fairly high order of generality, and in order to investigate whether or not the observed behaviour is consistent with these hypotheses, lower-order hypotheses must be derived referring to directly observable and measurable phenomena.[13] The variables are either expressed by industry/environment, in which case subscript j is used, or by size class in a particular industry, in which case subscripts ij are used. ($i = 1, 2$; the two size classes are 1000–5000 employees and 5000 +)

(a) *Research and Development Expenditure*

Company-financed research and development is calculated by:

$$X_{ij} = \frac{R_{ij}}{S_{ij} V_j} \times 100\%$$

where X_{ij} is an estimate of company expenditure in the ith size class in the jth industry as a proportion of value added. R_{ij} is company funds for R & D performance, by industry and size of company for 1963 (from N.S.F., 1966, table A-15, p. 93). S_{ij} is net sales of R & D performing companies by industry and size of company[14] for 1963 (from N.S.F., 1966, table A-36, p. 111). R_{ij}/S_{ij} measures research intensity as a proportion of net sales in R & D performing companies (see column (1), Appendix I). However, this tends to understate research intensity since non-resource costs have to be covered by company sales; value added would be a better measure of company size, although the N.S.F. figures are given only for net sales. Consequently, V_j is used as an adjuster to give a better estimate of research intensity as a measure of use of resources. V_j is value added by manufacture divided value of shipments for the appropriate S.I.C. grouping in the 1963 census of manufactures[15] (from Bureau of the Census, 1970, pp. 701–5).[16] V_j is measured for all firms in an S.I.C. grouping, and consequently the number of firms surveyed may be more than the R & D performing firms in some industries. Further, V_j is not available for the N.S.F. size categories, $i = 1, 2$.[17] Assuming V_j does not vary substantially between R & D and non-R & D performing firms in an industry, and that it does not vary substantially from size categories 1 and 2 in the N.S.F. survey, it may be used to adjust research intensity into an estimated percentage of value added by R & D performing firms, in the ith size category in the jth industry.[18] It was felt essential to incorporate some value added adjuster in the analysis, since the theoretical formulation is with respect to allocation of corporate *resources*: value added as a proportion of sales varies widely between industries owing primarily to technological differences (see Appendix I) and consequently a sales-based empirical analysis would be of little, if any, use.

We also investigate determinants of total funds for industrial research and development for the ith class in the jth industry, Y_{ij}:

$$Y_{ij} = \frac{T_{ij}}{S_{ij} V_j} \times 100\,\%$$

where T_{ij} is total funds (company and federal) for R & D performance, by industry and size of company for 1963 (from N.S.F., 1966, table A-3, p. 82).

(b) *Basic Research Expenditure*

Basic research expenditure, B_{ij}, is measured as a percentage of total funds for research and development by industry and size of company. Funds for basic research performance by industry and size of company for 1963 are provided in N.S.F. (1966, table A-6, p. 140), while the corresponding figure for total R & D performance is measured by R_{ij} (table A-3, p. 82).

(c) *Basic Research Performance*

The variable N_{ij} is used to measure the percentage of R & D performing firms in the ith size class in the jth industry conducting basic research in 1963. The data are provided in N.S.F. (1966, table A-73, p. 143).

(d) *Technological Opportunity*

This is defined by Phillips (1966) as 'the extent to which current science permits functional (as contrasted with stylistic) product changes and product differentiation among firms' (p. 305). Phillips obtained his measures of technological opportunity by studying the descriptions of the products primary to each four-digit S.I.C. industry included in the two-digit groups. The figure for each industry is the modal number for each industry. Since P_j operates as an environmental variable, its value is assumed to be the same for all firms in a particular industry in the N.S.F. classification scheme.

(e) *Growth*

The Bureau of the Census (1968b) produces indices of production to provide measures of the change in the value of work done in establishments in each S.I.C. grouping, valued in constant dollars to eliminate the effects of price changes. These are carried out for each four-digit sub-industry, and aggregated through to give values for each three- and two-digit industry. Current dollar weight data are provided for each two-, three- and four-digit classification. The change in value of work done for each S.I.C. classification from 1958 to 1963 was chosen as the relevant growth variable. Growth figures were available for the specific N.S.F. grouping in a number of cases; where it was not, it was obtained from constituent three- and four-digit industries using current dollar weight data and 1958 value added to weights. These are reproduced in column (6) of Appendix I as a proportion of the 1958 values and denoted by G_j.

Past growth is used as the independent variable, since, as was stated earlier, it is the historic effect of variables on resource allocations that affect the preference system for resources. Again, the variable is interpretable as an environmental one, since industrial growth as a whole may signal opportunities for innovation, even if individual firms are relatively slow-growing.

(f) *Federal Funds for Research and Development*

The variable F_{ij} is used to measure federal involvement in R & D in the ith size category in the jth industry and is measured by R_{ij}/T_{ij}. As F_{ij} increases from 0 to 1, federal involvement in R & D decreases as a proportion of all funds for R & D. While we would expect federal involvement to affect firms individually, and operate as an intra-firm variable, the highest level of disaggregation available in this analysis is at the ith size class in the jth industry.

(g) *Q.S.E.s as Proportion of R & D Expenditure*

This was calculated as total wages and salaries of scientists and engineers as a percentage of total R & D costs for the jth industry (N.S.F., 1966, table A-19, p. 97), and denoted as U_j.

(h) *Size of Research Establishment and Firm*

An average value is calculated for both variables, two measures being estimated for size of research establishment. In each case n_{ij} was used, where n_{ij} is the number of R & D performing companies in the ith size category in jth industry:

$$E_{ij} = \frac{R_{ij}}{n_{ij}} \quad \text{(millions of dollars)}$$

$$E'_{ij} = \frac{T_{ij}}{n_{ij}} \quad \text{(millions of dollars)}$$

$$Z_{ij} = \frac{S_{ij} V_j}{n_{ij}} \quad \text{(millions of dollars)}$$

Therefore, E_{ij} is the average size of research establishment measured in company funds for R & D, by size of company and industry, while E'_{ij} is average size of research establishment measured in total funds for R & D. Since the relationship between federal and company funding is uncertain, the two different measures of size of research establishment will be used in the regression analysis. Average size of firm is an estimate of average value added in the ith size category in the jth industry for the average firm measured by Z_{ij}.

LEVEL OF DATA AGGREGATION AND FUNCTIONAL FORMS OF THE REGRESSION EQUATIONS

We now have a set of operational variables derived from the earlier resource-based hypotheses. However, these are all measured at relatively high levels of aggregation, either at the level of the jth industry, or the broad ith size category in the jth industry. While we would wish for measures of P_j and G_j at this level of aggregation anyway, for the rest of the variables these measures must be judged inferior to enterprise-specific measures. Further restrictive assumptions must be employed if regression analysis is to be utilised.

The simple device of assuming that the *ij* estimates pertain to a 'representative firm' will be employed. This avoids and obscures the problem of variability within the *i*th size category in each industry, but it permits the variables developed above to be used in the following regressions. However, while a comprehensive analysis would incorporate enterprise-specific variables, our intention here is mainly to analyse reasons for differences between groupings of firms and examine to what extent they are attributable to differences in conditions between industries. While federal funding of R & D and size of firms are interpretable as intra-firm variables, the main burden of our analysis is concerned with inter-industry differences and the possible effects of environmental or industry-level variables. In the next chapter we shall present some evidence to suggest that the effect of industry-level variables is typically more important than that of intra-firm variables, especially when firms from a particular industry are sampled over restricted size ranges, as in our analysis. The evidence suggests that variability in innovative activity between industries tends to be much more important than variability within industries, for large firms. This would mean that the relative neglect of intra-firm variables in our analysis may be justified in terms of the relative importance of environmental and intra-firm variables in determining R & D activity.

A point in favour of grouping the firms at the two-digit level is Nelson's diversification hypothesis. As Nelson et al. (1967) point out, 'organised R & D is predominantly carried on by large firms, which also have more diverse product lines' (p. 53). In those circumstances, analysis at the four-digit level would be inappropriate since, if the Nelson hypothesis is correct, organised R & D must be related to firms operating at a considerably more diverse level of operations, possibly at three-digit, two-digit, or even higher levels of diversification.

As far as the functional form of the regression equations is concerned, four different forms are tried. The main burden of the hypotheses falls on directional effects of independent variables, and *a priori* there is no specific functional form which might be expected to be appropriate to the exclusion of others. However, there are reasons for the choice of different forms, and these are discussed below.

(1) The linear form (Fig. 7.7) is the simplest form to be fitted, and this is sufficient to justify its inclusion. If we had no reason to suspect the

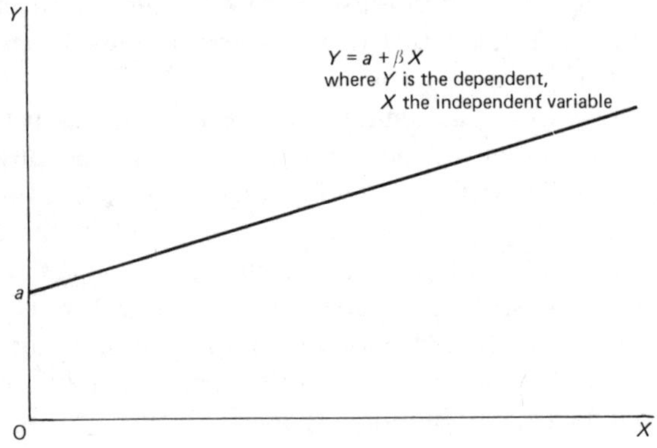

FIG. 7.7 Formulation (1): linear form ($Y = \alpha + \beta X$, where Y is the dependent and X the independent variable)

existence of non-linearities and saturation levels, this form would be sufficient to test the basic hypotheses below. However, there are reasons why the linear form may be an inferior specification of the relationship between research activity and the independent variables, owing to the violation of the basic assumption of linearity. As specified above with the inclusion of the variable α, the form provides a test of the reasonableness of the assumptions of strict convexity, since a significantly positive or negative α would be consistent with the existence of corner solutions.

(2) The double-log transformation (doub. log. transform) (Fig. 7.8) is in one respect a preferable specification of the functional relationships compared to the linear form, since it is consistent with the assumption of strict convexity. It is also a form which provides one test of non-linearity in the relationship between Y and X.

(3) The reciprocal transformation (recip. transform.) (Fig. 7.9), like the linear form, provides for the existence of threshold effects, violating the axiom of strict convexity. An important feature of this form is the existence of an asymptotic level which research activity tends towards. There is a theoretical limit to R & D and basic research activity in each case of 100 per cent; however, it is likely that in practice there is an asymptotic level operational substantially less than 100 per cent for both the R & D and basic research subsystems. Provided research activity in some cases is operating around this saturation level, the recip. transform. is one form of the regression equation which would test for such activity.

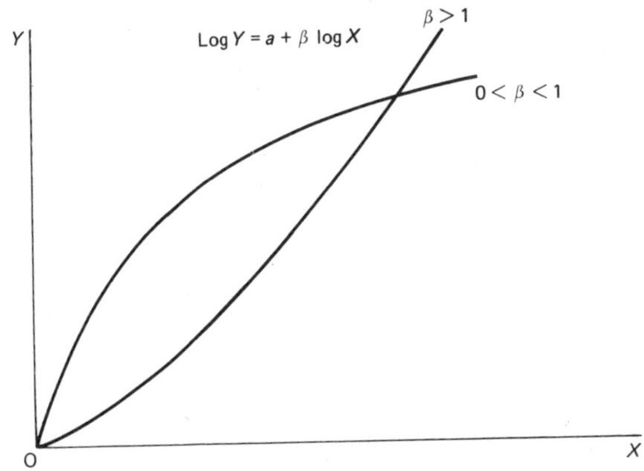

FIG. 7.8 Formulation (2): double-log transformation ($\log Y = \alpha + \beta \log X$)

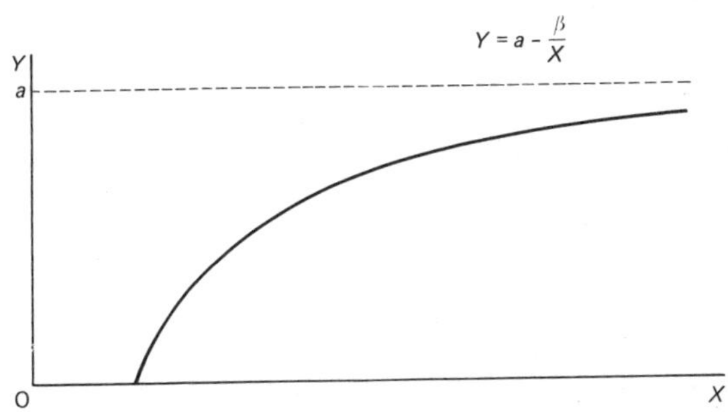

FIG. 7.9 Formulation (3): reciprocal transformation ($Y = \alpha - (\beta/X)$)

(4) The S-shaped curve of the logarithmic-reciprocal transformation (log. recip. transform.) (Fig. 7.10) has the properties of being consistent with the assumption of strict convexity and the *a priori* expectation of an asymptotic level for the appropriate research activity.

Thus, the different functional forms provide alternative formats for the testing of the hypotheses, each formulation investigating different possibilities as to the type of relationship between the independent variables and the appropriate research activity.[19] Prais and Houthakker (1955), in a study of consumer behaviour, utilised similar formulations on the grounds that it was only possible to test the reasonableness of a hypothesis against a limited number of specified alternatives (p. 88). Our

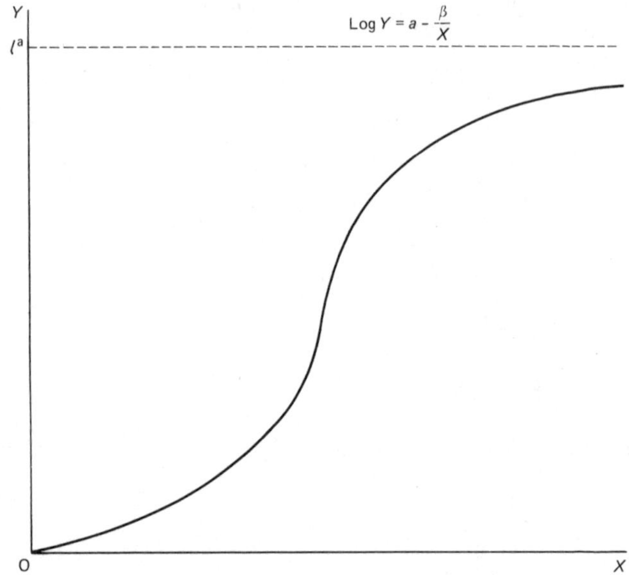

FIG. 7.10 Formulation (4): logarithmic-reciprocal transformation (log $Y = \alpha - (\beta/X)$)

argument here is similar since the different formulations are consistent with different aspects of the possible relationships between variables: linearity (formulation (1))/non-linearity ((2), (3) and (4)), strict convexity ((2) and (4))/threshold effects ((1) and (3)), saturation levels ((3) and (4)). In the subsequent analysis we shall utilise each formulation to examine the reasonableness of the alternative specifications of relationships between variables.

TESTING THE X_{ij} HYPOTHESES

In the regression equations of the remaining sections of this chapter, the value shown under the coefficient of each independent variable is the *t* value for that coefficient; if the value is significant at the 95 per cent level of confidence (two-tailed test) it is asterisked (*); if significant at the 99 per cent level this is indicated by a double asterisk (**). For each equation, *F* values, R^2 and $\overline{R}^2$ are also provided, and the sample which the regression uses is also indicated. From the sample code the composition of the sample may be examined by reference to the samples matrix of Appendix II. A further convention adopted is that in discussing a variable and its various transformations, in general only the code for the untransformed variable is given; for example, in discussing the performance of the technological opportunity variable in the different functional forms, we generally use P_j as signifying technological opportunity in untransformed or transformed form.

As an initial examination of the possible determinants of X_{ij}, company-financed R & D for the *i*th size class in the *j*th industry, a pilot regression was run using P_j (technological opportunity), G_j (industry growth), U_j (Q.S.E. intensity) and Z_{ij} (average size of firm in the *i*th size category). The only industries included in the sample for this equation are those for which Phillips originally provided estimates of P_j and for which N.S.F. data for 1963 are available. The simple linear form was used for this regression and the results are shown in Table 7.1.

The fit of the regression equation is quite good, with an F value of 14.20 and and $\bar{R}^2$ of 0.76. As far as the individual variables are concerned, both P_j and G_j are significant, P_j at the 99 per cent level, and both variables in the direction hypothesised. An interesting feature of the dual significance of P_j and G_j is that it indicates that, as far as this sample is concerned, fears that multicollinearity of P_j and G_j might make it impossible to separate out their respective influences on X_{ij} were unfounded. Both variables may therefore be retained in the analysis.

A difficulty with using only Phillips' values in the data matrix (column (5), Appendix I) is that it limits the degrees of freedom available. The device of utilising values for the *i*th size category in each industry is one means of extending the available degrees of freedom, but if further estimates of P_j were available, the numbers sampled could be extended quite substantially. In this respect, Dr Frank Swenson of the Department of Industrial Science, University of Stirling, provided invaluable assistance by supplying independent estimates for the remaining industries and sub-industries in the N.S.F. classification.[20] These are indicated by encircling in column (5), Appendix I.

In the second run, the provision of extra estimates of P_j permitted extra industries and sub-industries to be included in the analysis, increasing the degrees of freedom available. The second run also extended the analysis by using each of the different functional forms suggested in the previous section, and by including the variable F_{ij} (proportion of industry R & D company funded), whose possible significance for R & D allocation was not fully appreciated at the time of the pilot run. The remaining difference from the pilot run is that U'_j was

TABLE 7.1

Eq. no.	*Dependent variable*	*Independent variables*					*F*	R^2	$\overline{R}^2$	*Sample*
	Corporate-financed R & D	*Constant*	*Technological opportunity*	*Industry growth*	*Q.S.E. intensity*	*Size of firm*				
(1.1)	X_{ij}	-10.41 -2.67*	$+1.84P_j$ 4.52**	$+6.65G_j$ 2.28*	$+0.029U_j$ 0.48	$+0.0043Z_{ij}$ 1.66	14.20	0.81	0.76	(a)

used instead of U_j ($U'_j = 100 - U_j$).[21] The resulting regression equations are summarised in Table 7.2.[22]

All equations in Table 7.2 are fairly good fits, and each significant variable has the hypothesised sign. In each case P_j, G_j and Z_{ij} are significant except for G_j in eq. (2.2). From the F values of each equation, the best estimating equation is the sigmoid shaped log. recip. transform. with a marginally higher F value than the doub. log. transform. Provisionally, therefore, the results support the log. recip. transform., and the P_j, G_j and Z_{ij} hypotheses. No support is found for the U'_j hypothesis.

The non-significance of the F_{ij} coefficient indicates that the results of the regressions are *not* consistent with the hypothesis that federal funds substitute company funds for R & D. To examine further the implications of this result, the regressions were run again with the same sample and independent variables, but using Y_{ij}, total funds for R & D in the *i*th size class in the *j*th industry, as dependent variable rather than X_{ij}.[23] If federal funds augment rather than substitute company funds for R & D, then Y_{ij} should be inversely related to F_{ij}; as F_{ij} decreases, the federal proportion in R & D funding increases, and so also should total funds for R & D. This is examined in the appropriate regression equations (3.1)–(3.4) inclusive (see Appendix III for full results). In this case the best estimating equation is the doub. log. transform. (see Table 7.3).

Both the P_j and F_{ij} variables are significant at the 99 per cent level of confidence. When eq. (3.2) is considered in conjunction with eq. (2.1)–2.4) inclusive, the inference to be drawn is that when F_{ij} is considered as a possible determinant of research activity along with other independent variables, the results are inconsistent with the widely held belief that federal funding of R & D merely replaces what would have been allocated as company resources for R & D.[24] Instead it is consistent with the observation by Collier (1963) that R & D resources tend to be hired when federal contracts are obtained, and laid off when the contract ends.[25] This result has obvious policy implications since it does not support the conventional criticism of federal funding, and indeed it suggests that federal funding of R & D does not significantly affect the health of company R & D in different industries, at least in terms of intensity of company R & D funding. Instead, it implies that assessment of federal financing of R & D need not, in general, take account of possible costs of sacrificed corporate R & D since it does *not* appear to inhibit privately financed R & D.

TABLE 7.2

Eq. no.	Dependent variable	Independent variables						F	R^2	$\bar{R}^2$	Sample
	Corporate-financed R & D	Constant	Technological opportunity	Industry growth	Federal funding	Size of firm	Q.S.E. intensity				
(2.1) linear	X_{ij}	−4.56	$+1.96P_j$	$+6.02G_j$	$-0.040F_{ij}$	$+0.0054Z_{ij}$	$-0.064U'_j$				
		−1.06	5.21**	2.91**	−0.024	2.60*	−1.08	18.78	0.81	0.77	(d)
(2.2) doub. log.	$\log X_{ij}$	−5.71	$+1.16 \log P_j$	$+1.38 \log G_j$	$+0.096 \log F_{ij}$	$+0.14 \log Z_{ij}$	$+1.23 \log U'_j$				
		−1.38	7.70**	1.73	0.96	2.42*	1.22	23.56	0.84	0.81	(d)
(2.3) recip.	X_{ij}	14.71	$+\frac{5.45}{P_j}$	$+\frac{10.72}{G_j}$	$+\frac{0.0028}{F_{ij}}$	$+\frac{34.80}{Z_{ij}}$	$-\frac{88.30}{U'_j}$				
		4.24**	−6.53**	−2.86**	−0.19	−2.91**	0.40	17.66	0.80	0.76	(d)
(2.4) log. recip.	$\log X_{ij}$	5.23	$+\frac{1.76}{P_j}$	$+\frac{2.24}{G_j}$	$+\frac{0.0020}{F_{ij}}$	$+\frac{8.46}{Z_{ij}}$	$+\frac{77.94}{U'_j}$				
		5.71**	−7.97**	−2.27*	−0.52	−2.68*	−1.32	24.96	0.85	0.82	(d)

TABLE 7.3

Eq. no.	Dependent variable	Independent variables						F	R^2	$\bar{R}^2$	sample
	Total Industry R & D	*Constant*	*Techno-logical opportunity*	*Industry growth*	*Federal funding*	*Size of firm*	*Q.S.E. intensity*				
(3.2)	log Y_{ij}	−9.49	+1.36 log P_j	+0.61 log G_j	−0.38 log F_{ij}	+0.12 log Z_{ij}	+2.20 log U'_j				
		2.12*	8.36**	0.70	3.48**	1.91	2.03	45.46	0.91	0.89	(d)

The regression equations (2.1)–(2.4) inclusive provide no support for the retention of the F_{ij} and U'_j variables. Consequently, these regressions were re-run with the omission of the F_{ij} and U'_j variables, and the results are summarised in Table 7.4.

The pattern of goodness of fit is generally the same as in eqs. (6.1)–(6.4), all equations providing good fits, with log. recip. again showing a marginally higher F value. All variables are significant in each formulation, P_j and G_j at the 99 per cent level in each case. On the available evidence, it appears that the regression analysis supports the P_j, G_j and Z_{ij} hypotheses, with the sigmoid log. recip. formulation providing the best-fitting regression equation.

Thus, the hypotheses that technological opportunity, growth and size of firm encourage research intensity are supported by the above analysis. However, the hypotheses that the Q.S.E. ratio in R & D expenditure and federal funding of industrial R & D are, respectively, positively and negatively related to corporate-financed R & D intensity are not supported by the equations. The non-significance of the Q.S.E./R & D ratio may be considered a refutation of the original hypothesis (that the 'human element' in R & D is the most significant resource), or it may simply indicate that the hypothesis was not satisfactorily operationalised or measured here.[26] However, as indicated above, the non-significance of the F_{ij} variable in eqs. (2.1)–(2.4) inclusive, when considered along with its performance in (3.2), provides both useful and surprising information with consequent policy implications.

There is an interesting symmetry in eqs. (6.1)–(6.4) inclusive. The resulting significant variables include one technological (opportunity) variable, one market (opportunity) variable and one variable measuring size of system. As far as the different formulations of the equations are concerned, there is little evidence to choose between formulations, all equations recording high and similar $\bar{R}^2$ values. However, as discussed earlier in the chapter, each functional form has different implications for interpretation of the behaviour of allocations to R & D in response to changes in the variables, and since there appears to be little to choose between the various functional forms in terms of goodness of fit, it may be worthwhile considering, in particular, the assumption implicit in the recip. and log. recip. transformations that there is a saturation level or asymptote in terms of percentage allocations to R & D.[27]

For the recip. transform. the asymptote takes the value of the α coefficient, which in eq. (2.3) has the value 14.71 (per cent), and in the run which omitted F_{ij} and U'_j has the value 15.91 (per cent), as indicated by eq. (6.3). Both t values are significant at the 99 per cent level. The

TABLE 7.4

Eq. no.	Dependent	Independent variables				F	R^2	$\bar{R}^2$	Sample
	Corporate-financed R & D	Constant	Technological opportunity	Industry growth	Size of firm				
(6.1)	X_{ij}	−8.06	$+1.87P_j$	$+5.55G_j$	$+0.0047Z_{ij}$				
		−3.69**	7.14**	3.36**	2.43*	30.94	0.79	0.77	(d)
(6.2)	$\log X_{ij}$	−0.69	$+1.14 \log P_j$	$+1.92 \log G_j$	$+0.13Z_{ij}$				
		2.36*	8.72**	3.00**	2.35*	39.36	0.83	0.81	(d)
(6.3)	X_{ij}	15.91	$+\frac{5.29}{P_j}$	$+\frac{10.73}{G_j}$	$+\frac{34.23}{Z_{ij}}$				
		6.81**	−7.04**	−3.36**	−3.04**	31.59	0.80	0.77	(d)
(6.4)	$\log X_{ij}$	4.40	$+\frac{1.82}{P_j}$	$+\frac{2.65}{G_j}$	$+\frac{8.14}{Z_{ij}}$				
		6.92**	−8.88**	−3.05**	−2.66*	41.42	0.84	0.82	(d)

implication is that the best estimate of a ceiling to R & D allocations is about 15 per cent of value added in this specification of the regression equation.

In the log. recip. transform. the asymptotes have the value e^{α}, which is 186.79 (per cent) in eq. (2.4) and 81.45 (per cent) in eq. (6.4), the first estimate being infeasible and the second non-plausible. A problem in measuring asymptote values in both recip. and log. recip. transformations is that the highest value of X_{ij} included in the sample is 5.5 per cent in the case of scientific instruments, whereas the imputed asymptotes derived from the best-fitting regression line are substantially higher; this is reflected in the confidence intervals surrounding both formulations, the range being 4.96 to 24.46 per cent for α in eq. (2.3) (recip. transform.) and 14.43 to 2146.31 per cent for e^{α} in eq. (2.4) (log. recip. transform.), both at the 99 per cent level. The width of the confidence intervals is a consequence of the availability of data points being restricted to only lower values of X_{ij} as far as the theoretically feasible range from zero to asymptote is concerned. The problem is exacerbated in the case of the log. recip. transform. since a logarithmic transformation here puts a higher weighting on lower values of the dependent variable. Consequently, while both eqs. (2.3) and (2.4) are good fits, the grouping of the data points at low values of X_{ij} means that very little can be usefully stated with respect to possible saturation levels.

Therefore, in this section, each of the functional forms used to analyse possible determinants of X_{ij} has provided good-fitting equations, with three of the five independent variables recording significant t values in each case. There is little to choose between the functional forms in terms of goodness of fit. In the next section we shall similarly analyse basic research activity in an examination of possible determinants of allocations to that function.

TESTING THE BASIC RESEARCH HYPOTHESES

The N.S.F. data were similarly used to calculate possible determinants of basic research activity (Table 7.5). As with the X_{ij} variable, a pilot run was made using the lower formulation and only P_j estimates provided by Phillips. In this case the F_{ij} variable was included in the pilot run since there were specific and unambiguous expectations as to the sign of F_{ij}, unlike in the X_{ij} pilot run. Both E'_{ij} and E_{ij} were used in alternate regressions to measure average size of research establishment. The dependent variable was B_{ij}, percentage of total funds for R & D

TABLE 7.5

Eq. no.	Dependent variable	Independent variables						F	R^2	$\bar{R}^2$	sample
	Basic research	Constant	Technological opportunity	Industry Growth	Size of research establishment	Federal funding	Q.S.E. intensity				
(1.2)	B_{ij}	−24.18	$+7.81P_j$	$-12.48G_j$	$+0.19E'_{ij}$	$+62.67F_{ij}$	$-0.70U_j$				
		−1.14	2.97*	−0.79	2.73*	3.44**	−2.55*	3.65	0.70	0.50	(b)
(1.3)	B_{ij}	5.39	$+1.13P_j$	$-0.97G_j$	$+0.52E_{ij}$	$+10.88F_{ij}$	$-0.37U_j$				
		0.28	0.44	−0.058	2.34*	1.02	−1.28	2.97	0.65	0.43	(b)

allocated to basic research for the ith size category in the jth industry.

The initial results are surprisingly good considering the few degrees of freedom available and the frequently alleged non-economic basis of basic research. Size of research establishment is significant in both cases, but when E'_{ij} is used, P_j, F_{ij} and U_j are also significant. However, U_j has the opposite sign to that hypothesised, indicating that a negative relationship exists in this sample between the proportion of R & D cost accounted for by wages and salaries of Q.S.E.s and B_{ij}. This is a result which is difficult to account for from the preceding hypothesis.[28] Otherwise the only other coefficient in eq. (1.2) whose behaviour is not consistent with the earlier hypotheses is that of G_j; in this case the β coefficient is statistically insignificant.

To test the possible significance of the relationships between these variables and B_{ij} more fully, the four functional forms were utilised in a series of regressions in which the extended estimates of P_j were used. Each of the variables hypothesised to affect B_{ij} was included in the regressions, the size of research establishment variable being measured in both E_{ij} and E'_{ij} forms. The results were extremely poor, only one variable in the eight regressions being significant at the 95 per cent level, E_{ij} in eq. (4.2) (see Appendix III for the complete results of the regressions; equation numbers are (4.1)–(4.7) inclusive). Equation (4.2) shown in Table 7.6 also had the highest F value.

Considering the number of variables tested and the number of regressions run, it would not be surprising if at least one variable recorded a significant t value even if there is no underlying relationship between the independent variable and B_{ij}'.[29] Consequently, the regression results cannot be cited in support of the behavioural hypothesis. A further set of regressions was run in which P_j, G_j and E_{ij} were regressed individually against B_{ij} in each functional form (eqs. (7.1)–(7.12) in Appendix III) in order that extra degrees of freedom could be made available in the regression equations. However, the results were again extremely poor, none of the variables recording a significant t value. The regression analysis does not support any of the earlier hypotheses as to possible relationships between the independent variables and B_{ij}.

TESTING THE SPILLOVER MODEL

The failure of the regression analysis to support the B_{ij} behavioural hypotheses could mean that the hypothesised relationship between B_{ij} and independent variables was incorrectly specified either in terms of the

TABLE 7.6

Eq. no.	Dependent	Independent variables						F	R^2	$\bar{R}^2$	Sample
	Basic research	Constant	Techno-logical opportunity	Industry growth	Size of research establ-ishment	Federal funding	Q.S.E. intensity				
(4.2)	B_{ij}	−33.30	$+0.73P_j$	$-1.57G_j$	$+0.43E_{ij}$	$+12.36F_{ij}$	$+0.40U'_j$				
linear		−1.32	0.47	−0.15	2.36*	2.00	1.31	2.15	0.47	0.25	(e)

effects of the variables or the appropriate functional relationship. A further possibility is that the available highly aggregated data is not a suitable form for the testing of hypotheses originally specified at the level of the individual firm.

Some indication as to how the present theoretical construction may be inappropriate has already been discussed in the section in Chapter 4 dealing with managerial preferences and R & D subsystems. There it was suggested that managers may rank R & D subsystems according to the extent to which each subsystem is removed from final output. Radicalness and uncertainty of subsystem activity with respect to their contribution to final output encouraged managerial bias and discrimination against such subsystem activity. As Gold (1971, p. 222) points out, in such circumstances management may possess a ranking system for subsystem activity, familiar and routine operations or activities. Such behaviour is described in Fig. 7.11(b).

Figure 7.11(b) indicates the direction in which resource preferences are typically ranked, from production to basic research, the preference ranking operating in the opposite direction to the chronological process implicit in the traditional linear model of innovation represented in Fig. 7.11(a). In conjunction with the concept of hierarchical arrangement of R & D subsystems (see Chapter 2) and managerial preference for familiar and minimally uncertain research activity, Fig 7.11(b) suggests the behavioural model of resource allocations in Fig. 7.11(c). The diagram is to be interpreted as implying the ranking of resource activity in Fig. 7.11(b); the first preference of management is for the R & D activity indicated by the chain *cdc*. In this case resources are allocated to development which in turn develops project ideas which in turn move activity back to the production stage. A lower-ranking preference is for the activity chain *bcdb* resulting from resource commitment to applied research, with the least preferred activity being *abcda*. The arrows operating in the opposite direction indicate the possibility that resources may not be allocated in a simple linear progression *abcd*, as Machlup (1962a) earlier pointed out. For example, a project which is referred back to basic research from applied research twice and development once might follow the flow process of research activity and resource commitment *abcababcd*.

Thus, while it is necessary for an activity at the 'top' end (end *a*) of the R & D spectrum to pass through all intervening stages between it and activity *d* (intervening stages are *b* and *c* here), a project may be referred back to any of the stages preceding it by 'skipping' intervening stages. This is indicated in Fig. 7.11(c) by the arrows bypassing *b* and *c*. It is of

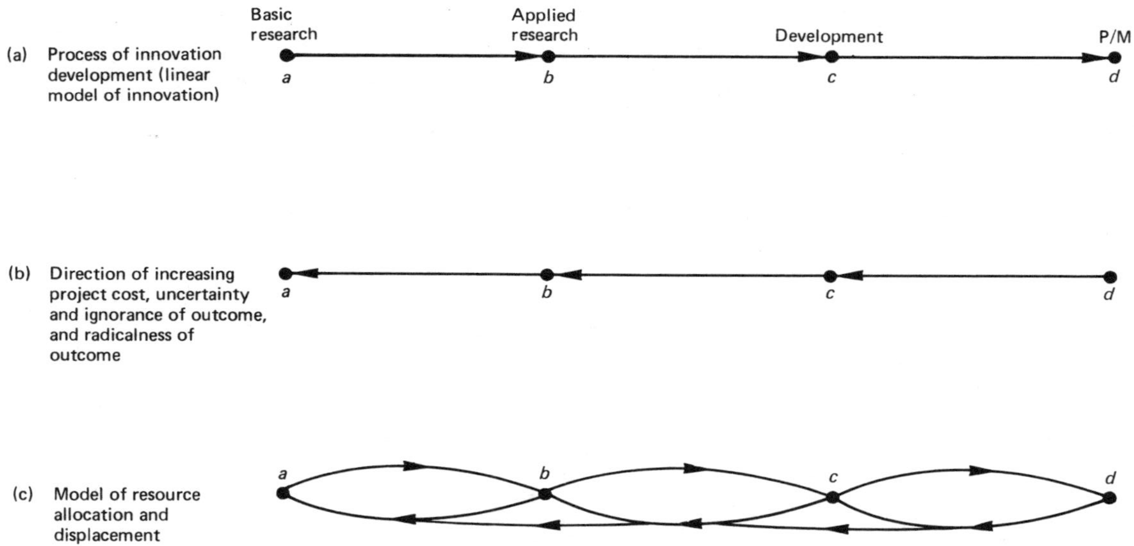

FIG. 7.11. (a) Process of innovation development (linear model of innovation). (b) Direction of increasing project cost, uncertainty and ignorance of outcome, and radicalness of outcome. (c) Model of resource allocation and displacement

course possible that complex R & D projects may involve activities being carried out simultaneously though at different stages and on different sections of the innovation. Such complex behaviour would have to be studied in a more sophisticated version of the model than is developed here.

The asymmetry of resource preferences within the R & D function[30] implies that there is a threshold to be overcome before a resource is committed to a particular category. Since the firm's first preference is to minimise displacement from P/M, the firm's primary operations and activities are concerned with P/M resources, with development, applied research and basic research being less preferred options, in that order. This suggests that the assumption of strict convexity utilised in the model of the managerial preferences system may be inappropriate in this context. Instead, Fig. 7.11(c) suggests a 'spillover' model of R & D subsystem allocations in which there is progressive commitment of resources along the R & D spectrum from P/M activity, the boundaries between subsystems representing thresholds which have to be crossed before a particular subsystem activity is undertaken and resource commitment is entered into. For example, a firm may allocate resources to the *cdc* 'loop' initially, with minor developments of existing products passing through into production, and new projects being created and conceived within the boundaries of this development–P/M loop. The preference system, in ranking subsystem activity according to uncertainty and radicalness of associated final output, creates thresholds or barriers which must be overcome before a lower-ranking activity is undertaken. As indicated in Fig. 7.11(b), the barriers are overcome in the order *dcba*. Thus, the development–P/M loop may first be extended to include applied research, and finally basic research as the barriers to adoption of specific subsystem activity are progressively overcome.

This raises the question as to whether the hypotheses developed earlier with respect to basic research activity may have any relevance to the present discussion. In fact a simple reformulation is to specify that the effect of changes in the respective variables is not primarily to reallocate resources within existing subsystems but instead is to strengthen or weaken the threshold boundaries constituting barriers to allocation of resources to less preferred subsystems. For example, in the spillover model the hypothesised effect of increased industrial growth is not to increase basic research share of R & D activity, but to weaken the threshold inhibiting the adoption of basic research activity in the first place. Similarly, the other variables affect subsystem resource allocation in an analogous manner to that of the earlier B_{ij} hypothesis, except that the effects of the variables are with respect to facilitating or inhibiting the triggering of a particular subsystem activity rather than with respect to reallocation of resources within the existing subsystems.

To test this formulation, the variable N_{ij} substituted the variable B_{ij}in the regression analysis. N_{ij} is the percentage of R & D performing companies in the *i*th size class in the *j*th industry who also conducted basic research in 1963. N_{ij} is therefore a measure of the proportion of R & D performing companies who have crossed the basic research threshold in a particular industry. Using N_{ij} is a less ambitious test of the model than B_{ij} since it is a simple measure of the extent to which basic research activity in an industry is switched 'on' or 'off' rather than the distributional aspects of subsystem activity measured by B_{ij}. However, recalling what was said earlier about the likelihood that subsystem stability diminishes further down the resources hierarchy, this may be the best that can be expected from analysis of basic research. The basic research subsystem is the most vulnerable to short-run distress conditions, as the spillover model would suggest, and it is also generally the most difficult to analyse in terms of anticipated resource utility. The non-significance of the B_{ij} equations could be a result of natural inherent instability and irregularity of allocations to this function rather than specification error in the regression analysis itself. Unreliability of the distributional aspects of subsystem activity notwithstanding, it may be possible to establish pattern in the relatively simple 'switch mechanism' implicit in the spillover model.

The N_{ij} regressions were run using the extended sample provided by the extra estimates of P_j, in the four forms with both E_{ij} and E'_{ij} measures of size of research establishment. The results are summarised in eqs. (5.1)–(5.8) inclusive in Appendix III. Both versions of the doub. log. transform. turned out to be good estimating equations of log N_{ij} and were unequivocably superior to the other estimating equations when F values were compared. Equation (5.3) in Table 7.7 recorded the highest F value.

As with the X_{ij} equations, F_{ij} and U'_{ij} perform poorly in eq. (5.3), and this pattern is repeated in the set of equations (5.1)–(5.8) inclusive. The only significant t value for either variable is for F_{ij} in eq. (5.4). The signs for G_j and E'_{ij} are those anticipated, but the sign of the regression coefficient for P_j is negative, indicating that the proportion of R & D performing companies in a particular industry undertaking basic research *diminishes* as technological opportunity increases, *ceteris paribus*. This implies the surprising conclusion that the proportion of firms conducting basic research is inversely related to the extent to which

TABLE 7.7

Eq. no.	Dependent variable	Independent variables						F	R^2	$\overline{R}^2$	Sample
	% conducting basic research	Constant	Technological opportunity	Industry growth	Size of research establishment	Federal funding	Q.S.E. intensity				
(5.3)	log N_{ij}	−0.004 −0.00055	−0.58 log P_j −2.59*	+3.26 log G_j 2.29*	+4.2 log E'_{ij} 5.38**	+0.080 log F_{ij} 0.51	+0.52 log U'_j 0.28	12.00	0.83	0.76	(f)

current science permits functional product changes, *after allowing for the effects of industrial growth and size of research establishment*. This is discussed further below.

Since F_{ij} and U'_j performed badly in these regressions, the four functional forms were again used to re-estimate determination of N_{ij} using only E'_{ij} as measure of size of research establishment and deleting F_{ij} and U'_j. To examine what effect deletion of P_j would have on the fit of the equations, this was done in a second run for each functional form. The results are shown in eqs. (8.1)–(8.8) in Appendix III. Both doub. log. transformations again recorded the highest F values, the equation with the highest value being (8.3) in which P_j was included as an independent variable (see Table 7.8). The F value of eq. (8.3) and the respective t values of variables are in each case higher than the corresponding equation (5.3) in which F_{ij} and U'_j were included. The F value alone indicates that (8.3) is a better estimator of log N_{ij} than (5.3).

One result worth emphasising at this point is the apparently strong relationship between average size of research establishment and percentage conducting basic research (t value of 7.27) after allowing for the effects of P_j and G_j. While it was stated earlier that little could be said about the possible determinants of B_{ij}, average size of research establishment was the variable recording significant t values in each of the linear formulations of the B_{ij} regression equations discussed earlier (see, again, eqs. (1.2), (1.3) and (4.2)). It is therefore worth emphasising the possible significance of this variable as a determinant of basic research expenditure, as far as future analysis is concerned, given its apparent importance relative to other variables in both N_{ij} and B_{ij} analyses.

Again, however, while G_j and E'_{ij} recorded the signs anticipated, P_j still had a negative coefficient. The significance of the t value for P_j suggests that the regression analysis is not only inconsistent with the hypothesis regarding the possible relationship between P_j and N_{ij}, but directly contradicts it; it seems that high technological opportunity does not encourage the crossing of the basic research threshold, but in fact appears to inhibit it. The general resource-based approach to the determination of N_{ij} appears to have resulted in good estimating equations in the cases utilising G_j and E'_{ij};[31] what is it about the P_j variable or hypothesis that results in such an emphatic contradiction of the P_j hypothesis?

Some relevant aspects of the problem may be illustrated if we consider the process by which the respective variables might operate on the respective subsystems and resource preferences, In particular, the G_j and E'_{ij} variables differ from P_j in that they do not describe differences

TABLE 7.8

Eq. no.	*Dependent variable*	*Independent variables*				*F*	R^2	$\overline{R}^2$	*Sample*
	% conducting basic research	*Constant*	*Technological opportunity*	*Industry growth*	*Size of research establishment*				
(8.3)	$\log N_{ij}$	2.09 7.56**	$-0.60 \log P_j$ −2.97*	$+3.68 \log G_j$ 3.81**	$+0.40 \log E'_{ij}$ 7.27**	22.73	0.83	0.79	(f)

inherent in R & D activity conducted in the respective industries, but instead differences in the conditions encountered by subsystem output. Both G_j and E'_{ij} are hypothesised to operate on the R & D system by reducing the effective barriers to implementing less preferred subsystem activity; increasing G_j and/or E'_{ij} progressively reduces barriers and resistance to implementing subsystem activity along the chain *dcba* in Fig. 7.11(b).

However, the variable P_j is of an entirely different nature to G_j and E'_{ij}; it is an intended measure of one aspect of R & D work, 'the extent to which current science permits functional . . . product changes and product differentiation'. Earlier it was suggested that this variable facilitated linkages between basic research and applied research, and this is indeed liable to be the case. However, there is another aspect of this variable which is only apparent once the behaviour of basic research activity is considered in a spillover frame of reference, that is, the effect of P_j on applied research resource utility. The applied research subsystem is concerned with the discovery of new scientific knowledge with specific commercial applications; the variable P_j refers to linkage between science and technology and certainly will have a direct effect on the resource utility of applied research. This has significant implications for basic research in a spillover context.

If resource utility of applied research rises owing to increase in P_j, this may inhibit rather than facilitate spillover into basic research. If firms are operating in a rich applied research environment, incentive to direct resources to more speculative basic research may be small. The conservative bias implicit in the spillover model suggests that high and demonstrable resource utility from applied research might lend weight to arguments against basic research allocations; why allocate to basic research when there is such potentially high resource utility to be derived from allocating to those areas where there are obvious links between scientific knowledge and new products?

However, the R & D performing corporation operating in a low P_j environment is liable to find its applied research subsystem relatively barren compared to high P_j industries. In such circumstances there may be little inhibition with respect to spillover into basic research since basic research is not diverting R & D resources from areas of high resource utility. A similar point is made by Williams (1961) in a normative context: 'If . . . researchers wisely choose basic problems because their applied work has run up against existing scientific knowledge and has become unproductively "empirical" then the wastage rate need not be high' (p. 26).

As with the behaviour of the F_{ij} variable in the X_{ij} equations, this suggests an interpretation which again runs counter to the conventional wisdom. It seems that the industries in which R & D performing firms are more liable to adopt speculative and highly uncertain basic research are the non-science-based, if the effects of industrial growth and size of research establishment on N_{ij} are separated out. The science-based industries are more liable to 'play safe' by exploiting applied research and excluding basic research, *ceteris paribus*. In this sense, the non-science-based industries may be said to be more progressive in their readiness to undertake basic research activity compared to the science-based industries.

This reinterpretation opens the resource-based analysis tested by the regression equations to the criticism that it is non-refutable; had the original P_j hypothesis been supported by a significant and positive t value for log P_j in eq. (8.3), the original P_j hypothesis would probably have been retained without question. It was only when a negative and significant t value was obtained that an alternative explanation was sought. The only justification offered for this is that it was the recording of an aberrant result for P_j which prompted a reconsideration of the role played by the respective variables in the spillover model and clarification of the distinction between the effects of G_j, E'_{ij} and that of P_j. Changes in G_j and E'_{ij} do not imply qualitative changes in the inherent properties or characteristics of composite resources in respective subsystems, but only the conditions affecting the exploitation of output of existing or marginal resources of the respective subsystems. As both G_j and E'_{ij} increase, we expect a progressive unfolding of resource allocations to the less preferred subsystems along the chain *dcba* in Fig. 7.11(b) as it becomes perceived worthwhile to allocate resources to the next subsystem. Examining the implications of P_j more fully, however, it is apparent that the direct impact of this variable is oriented towards the applied research subsystem. It is this specificity of effect of P_j that encouraged the reassessment of its role in the spillover model, and it is this which we argue justifies using the alternative hypothesis, in which light the relationship between N_{ij} and P_j was reinterpreted.

Thus, the regression analysis supports the hypotheses that technological opportunity, growth and size of research establishment all affect the propensity to undertake basic research. There is a parallel here with the resulting significant variables in the equations used to examine possible determinants of company-financed R & D, in that similarly we obtain one technological (opportunity) variable, one market (opportunity) variable, and one variable measuring size of the subsystem. However, in this case the technological opportunity variable appears to be negatively related to the dependent variable. Tentatively, therefore, it appears that some aspects of basic research activity may be accounted for in terms of hypotheses developed in this chapter.

CONCLUSIONS

In this chapter the resource-based model of Chapter 5 has been utilised to formulate and test hypotheses concerning the conduct of R & D and basic research activity in industry. The results were generally quite good in terms of the fit of the regression equations and significance of variables, except in the case of the B_{ij} variable. As well as providing evidence which is of use in selecting which of the respective hypotheses is consistent with the observed behaviour, the multiple regression analysis also indicates some features of the distribution of R & D activity in the United States in 1963 not immediately obvious from casual observation. Formulating the regression hypotheses in the context of the resources model illuminates the postulated relationships and determinants involved in the technological change systems, and the resulting conclusions hopefully provide a clearer understanding of the important variables operating in the R & D decision-making process. On a more general level, it also appears that basic research activity may be explained to some extent in terms of the hypotheses developed earlier in the chapter. In the context of the frequently alleged non-economic basis and motivation of such research discussed earlier (see Chapter 2), this may be regarded as noteworthy in its own right.

However, there is another aspect to this chapter which should be stressed. As has been emphasised in earlier chapters, firms cannot optimise allocations through analysis and control at project level owing to the high degree of complexity and uncertainty surrounding projects. Consequently, neoclassical theory is generally inapplicable at this level, and at aggregative levels. On the other hand, the hypotheses developed here are framed in terms of the resources model of Chapter 5 which does not require analysis of component projects in order to provide a basis for decisions on overall resource aolocations to functions.

In placing emphasis on the formation of 'Gestalts' and perception of pattern in corporate/environment relations, the resources model stresses learning and adaptation in corporate allocations, which does appear to be an important aspect of corporate behaviour and decision-making at this level in the firm. Not only does this facilitate explanation of the 'top-down' nature of allocations in the corporate hierarchy, but it also permits explicit recognition of the significance of uncertainty at lower levels in the corporation; since analysis of higher levels is not equivalent to analysis of aggregates of lower-level projects, we do not require in the first instance deterministic or stochastic project modelling before we can study higher levels.

Consequently, the resources framework provides a basis on which models of *rational* decision-making may be introduced in analysing corporate behaviour in areas where rationality is not generally recognised. The previous studies which have used a rationality postulate, such as neoclassical approaches, have typically been based on shaky, project-level, foundations. It is hoped that the contribution of this chapter is demonstrated in the potential usefulness of an alternative basis for economic model-building, as well as in having provided some interesting results with possible implications for policy-making.

APPENDIX

The purpose of this appendix is to provide a brief survey of some of the more important studies of the determinants of innovative activity in the large modern corporation that are not discussed in the main body of the text. Hopefully, this will place the analysis of Chapter 7 in context. This was not done in the main text of this chapter since the analyses surveyed here tend to be based on a neoclassical framework, either explicitly or implicitly. Consequently, the selection, interpretation and emphasis of variables deemed relevant to the analysis is rather different from that of the systems framework. Where such differences arise and are important will be emphasised below.

We shall restrict consideration to the major hypothesised relationships discussed and analysed in the literature. The first of these is the question of possible relationship between size of firm and innovative activity.

(a) *Size of Firm*

This factor has been extensively investigated by a number of analysts using measures of size, in particular, assets, sales and employment. A stimulus for such study has been the Schumpeterian thesis (Schumpeter, 1954) that size and market power facilitate the process of dynamic competition through innovation. Hamberg (1966) distributed a sample of 340 firms taken from the *Fortune* 500 into 17 industries and found the elasticity of R & D effort (measured in terms of ratio of R & D employment to total employment) to size to be only weakly related to size, measured in terms of total employment and total assets. Comanor (1967), on the other hand, conducted an analysis similar to Hamberg's of 387 firms in 21 industries using late 1950s data and found the elasticity of research effort relative to total size (measured in employment terms) to be significantly less than one for 7 industries, and in no case significantly greater than one. Comanor's findings are consistent with Scherer's (1965) analysis of 352 firms from the *Fortune* 500 for 1955, in which he found that the largest firms accounted for a substantially smaller share of R & D employment than they did of sales. Mansfield (1968b), using 1945 data, found that the largest firms in the petroleum, drug and glass industries spent significantly less on R & D relative to sales than smaller firms, the chemical industry being an exception. Grabowski (1968) supported Mansfield's findings in an analysis of large corporations' allocations to research for 1959–62 in the chemical, drug and petroleum industries. Both Worley (1961) and Smith and Creamer (1968) found a tendency for firms in the middle of the size distribution of their respective samples to be more research-intensive than larger and smaller firms, in terms of employment and expenditure respectively.

As Scherer (1970, p. 361) comments, the evidence suggests that, *up to a point*, size leads to proportionately increasing innovative activity in most industries. However, beyond that point further size does not lead to increased innovative intensity, and may lead to less. This is consistent with the earlier resource-based hypothesis if, to the hypothesis that R & D advantages are gained from increasing size, is added the qualification that they may encounter a saturation point. Further than this, comparison with the study conducted in Chapter 7 is difficult to the extent that the studies conducted above are typically conducted on an individual industry basis, whereas the analysis of Chapter 7 is on an industry-wide basis.[32]

(b) *Concentration*

Measures of concentration of industry sales have been used as indicators of monopoly power to examine the possibility that this may have an effect on innovative activity, frequently with the Schumpeterian hypothesis in mind. Using 1947 and 1951–2 data, Horowitz (1962) found a weak association between concentration and intensity of research expenditure, as did Hamberg in his 1966 study. Scherer (1967), in an analysis of the determinants of intensity of research employment, initially found a strong relationship between the concentration index and research effort, but a much weaker relationship when differences in technological opportunity were allowed for. Comanor in his 1967 study found no significant relation between concentration and research intensity. Recent studies by Adams (1970), Philips (1971), Globerman (1973) and Howe and McFetridge (1976) report mixed results in each case with concentration being significantly related to research effort according to part of the evidence of each study, and uncorrelated according to other analysis in the same studies.

The evidence suggests, at best, a weak association between concentration and research intensity,[33] especially after allowing for technological opportunity which is typically strongly related to degree of concentration. It is not considered as a possible determinant of research intensity in the analysis of this chapter since we are using a non-discretionary model of corporate behaviour in which the assumption of hostile or potentially hostile environments ensures that market structure does not play an important role in deciding the distribution of resources. The available evidence suggests that on empirical grounds also its omission is justified.[34]

(c) *Organisational Slack*

Availability of cash and resources surplus to that immediately required to maintain the corporation has been frequently suggested as a source of R & D activity, a hypothesis which might be regarded as consistent with both Schumpeter's (1954) and Penrose's (1959) theories of corporate growth and development. Since depreciation may act as a source of liquidity, it has been used as a test of this hypothesis. Grabowski in his 1968 study found that a variable measuring after-tax profits plus depreciation was positively and significantly related to research effort in each of the three industries studied. However, Scherer in his 1965 study, Hamberg in his 1966 study, Mueller (1967) in an econometric analysis of 67 firms using late-1950s data, and Smyth et al. (1972) in a study of patenting intensity in U. K. industry, all found little or no relation between measures of profits and/or liquidity and research effort. Elliott (1971), in analysing research intensity for 53 firms over the period 1953–66, found that profit expectations were more important determinants of research expenditure than cash flow variables. Minasian (1962) in a study of 18 chemical firms for 1947–57 found lagged profit explained much better by R & D than lagged R & D was by profits. Branch (1974), however, found some evidence of a significant relationship between past profitability and research effort in a study of 111 firms in 7 industries for 1950–65.

Regarded together, the available evidence provides little evidence of a consistent relationship between measures of profit and liquidity, and R & D intensity. Since profits and cash surplus to operating requirements would facilitate discretionary behaviour, such variables are not included in the analysis of Chapter 7, and again we may note that the omission of such variables does not appear to be a severe defect, judging by the available empirical evidence.

(d) *Diversification*

Following Nelson (1959a), a few studies have explored the possibility that there may be some relationship between product diversification and research effort. Nelson argued that, in view of the inherent uncertainty with respect to final output of research activity (particularly basic research), the firm which has a variety of products and markets is more liable to be in a favourable position to exploit the unexpected opportunities thrown up by research, than is a more specialised firm. Grabowski in his 1968 study in fact found that degree of product diversification was related to research intensity after taking account of other factors. However, Comanor (1965), using late-1950s data for 57 pharmaceutical firms, found a negative relationship between degree of diversification and research output. Scherer, in his 1965 study, obtained mixed results in relating degree of product diversification to R & D employment.

As with the previous two variables discussed, concentration and organisational slack, the evidence is, at best, weak as far as possible relation to R & D effort is concerned. However, unlike the tests of the other two variables, the failure of past empirical analysis has direct relevance to the present analysis, since Nelson's diversification hypothesis can be simply expressed in resource utility terms, and related to non-discretionary steady-state behaviour. As mentioned earlier in this chapter, it has similarities to the hypothesis that basic research activity is directly related to size of research establishment. There is, however, no obvious reason for the ambiguous results obtained from the studies to date.

SUMMARY

This brief survey of empirical analyses of the determinants of R & D is perhaps best characterised by the lack of conclusiveness or finality as far as the relationship of any of the variables to R & D effort is concerned. We have concentrated on the major hypotheses which have been suggested and empirically examined in the literature; yet, with the possible exception of the relationship between size of firm and research effort, there is little definite which can be stated with respect to the determination of research effort, without heavy qualification. In this context, it is perhaps interesting that of the three most extensively analysed relationships, the two involving concentration and organisational slack are based on the supposed existence of managerial power and discretion, and consequently do not have a place in the analysis of this chapter. Despite the unequivocal support for the hypothesis that R & D effort is directly related to technological opportunity provided by Phillips (1966), Scherer (1967) and Comanor (1967), the emphasis in empirical research does not appear to have shifted from further analysing the possible empirical implications of market distortions or imperfections in the neoclassical frame of reference, nor is there any evidence of recent attempts to develop more sophisticated measures of technological opportunity, a variable difficult to measure, but apparently a fundamental determinant of research activity.

Consequently, this survey highlights the different orientation and emphasis between the systems frame of reference developed here, and the general concern of previous empirical studies. Tentatively, it is hoped that the framework developed here shows some promise as a basis for further empirical analysis of corporate behaviour. It is hoped to demonstrate a further application of the framework in the next chapter

8 Rivalry, Learning and Variation in Innovative Activity

In the last chapter we looked at inter-industry differences in R & D activity in the context of the open system framework. In order to do so, certain restrictive assumptions were placed on the analysis, and possible effects of intra-firm ϕ variables on the analysis were ignored. In this chapter we examine some aspects of variation in innovative activity[1] *within* industries. While the open system framework appears to have facilitated interpretation and analysis of R & D activity at industry level in Chapter 7, the assumption of 'representative firm' does ignore the fact that industries typically incorporate a rich variety of types of corporation operating with a wide range of values of the ϕ variables. While the representative firm argument is useful as a first analysis and for explanation of certain gross features of industry behaviour, we are also interested in the possibility that the open system framework may be relevant for studying internal aspects of industry behaviour in this area. In doing so, it will be useful to bear in mind the central assumption of Chapter 6, that management seek to establish a pattern in their perception of corporate–industry relations. The way that this resource preference system is built up is through feedback from experience of past allocations. The steady-state allocations are learnt through this iterative process of resource allocation.

In this chapter it is suggested that this interpretation provides a reconciliation of apparently conflicting evidence contained in three empirical studies by Grabowski (1968), Grabowski and Baxter (1973) and Scherer (1965). A recent study by Howe and McFetridge (1976) is also discussed, and its relationship to the earlier studies assessed. Firstly, however, it will be helpful to consider the analytical context in which these approaches were developed, that of rivalry in R & D work.

RIVALRY AND THE IMITATIVE HYPOTHESIS

Economic analysis of innovative decision-making in the firm has traditionally emphasised the significance of monopolistic and oligopolistic market structures in this area, with associated implications for model-building. Kamien and Schwartz (1975) confirm the central importance of oligopolistic rivalry in studies of technological competition:

> Efforts have proceeded along two routes to bridge the gap between traditional micro-economic models of competition and Schumpeter's model.[2] Both lines of work have focussed on the role of R & D rivalry in determining the pace of inventive activity and have utilised findings of previous empirical studies to guide assumptions and check conclusions. In the first group . . . R & D rivalry is supposedly confined among existing members of an industry who view each other with a Cournot oligopoly framework.[3] In the second set . . . , the emphasis is on potential rivalry from any quarter, as stressed by Schumpeter, and requires extension of the model along lines analogous to recent advances in the theory of limit pricing. (pp. 28–8)

The emphasis on imitation is explained by Grabowski (1968) in discussion of National Science Foundation interview studies (N.S.F., 1956):

> one strong trend of thought running through these studies is that firm decisions on R & D are strongly influenced by the behaviour of competitors, and, in particular, that a great deal of imitation exists among firms with respect to R & D allocations. Since most R & D is performed by firms operating on oligopolistic market structures and it is an activity presumably involving greater uncertainty than other undertakings, firms may imitate each other as a conservative strategy for minimising risks. (pp. 296–7)

Consequently, the empirical studies of innovative activity discussed in this chapter (Grabowski, 1968; Grabowski and Baxter, 1973; Scherer, 1965) developed their analyses in the context of technological competition regarded as a feature of oligopolistic market structures. In these studies, consideration is given to the possibility that variation in such activity within industries may be explained by variation in the propensity to 'match' or imitate competitors' allocations in innovative activity, particularly when measured as a percentage of sales. It is this last aspect which will concern us in this chapter and consequently it is appropriate at this point to discuss in some detail the hypothesised role played by oligopolistic competition in each of these studies.

Firstly, Scherer (1965) regressed patent output on sales for 14 two- and three-digit U. S. manufacturing industries using firms from the *Fortune* 500[4] list for 1955 (patents were lagged by four years to allow for the standard time necessary to house a patent application). The results are shown in Table 8.1. Scherer found a relationship between the industry regression coefficient and the R^2 for the industry equations in that the slopes of the regression equations were positively correlated with the R^2 for these equations with a rank correlation coefficient of 0.69. According to Scherer:

> The higher an industry's average patent output per sales dollar is, the less variable patenting tends to be relative to size. My interpretation of this result is that in technically progressive fields like electrical equipment and general chemicals, technological competition forces firms to match each other's inventive efforts. But in unprogressive fields like textiles, food products, paper and fertilisers, soap and cosmetics, invention is only a business option. (p. 1100n.).

Secondly, Grabowski and Baxter (1973) also found a similar relationship between the mean employment of R & D professional employees per million dollar sales and the coefficient of variation (measured by standard deviation ÷ mean) for 29 three-digit U.S. manufacturing industries for 1956 (see Table 8.2). Holding the level of industry concentration constant, the partial Kendall rank correlation coefficient between mean industry research intensity and the coefficient of variation in research intensity was −0.35 (significant at the 95 per cent confidence level). Grabowski and Baxter interpret this as signifying: 'As the decision-making environment shifts to one in which R & D becomes more important as a competitive weapon, the pressures and incentives for firms to react strongly to competitors' actions correspondingly increase. Hence the smaller observed coefficient of variation in the more research intensive industries, (p. 233).

Thus, as industry research intensity increases, variability in R & D employment tends to decrease. As with Scherer's findings, firms appear to react to increased competition in technological change by moving towards a common solution. Again, a tendency towards 'competitive matching' or imitative behaviour in technologically progressive in-

TABLE 8.1 Linear regressions of patenting on sales, by industry

Industry	*Intercept*	*Regression coefficient*	*N*	R^2	*Total patents*
Food and tobacco products	−0.4	+18.05 (2.04)	75	0.52	366
Textiles and apparel	2.8	−0.48 (7.00)	25	0.00	70
Paper and allied products	4.5	+7.11 (6.14)	21	0.07	120
General chemicals*	8.1	+262.48 (25.68)	41	0.73	3,316
Misc. chemicals†	13.0	+19.33 (20.50)	14	0.07	231
Petroleum	4.5	+81.10 (10.50)	30	0.68	2,194
Rubber products	7.3	+52.32 (11.29)	8	0.78	303
Stone, clay and glass	−12.4	+200.92 (25.03)	19	0.79	434
Primary metals	0.4	+23.21 (2.50)	50	0.64	486
Fabricated metal products and miscellaneous‡	5.9	+61.86 (16.56)	31	0.32	516
Machinery	6.1	+90.40 (12.58)	46	0.54	967
Electrical equipment and communications	22.5	+311.06 (17.61)	35	0.90	5,036
Transportation equipment, except aircraft	2.7	+59.72 (5.28)	30	0.82	1,685
Aircraft and parts	6.8	+70.38 (22.39)	23	0.32	739
All industries combined	10.7	+73.81 (4.09)	448	0.42	16,463

* Includes S.I.C. 281, 282 and 283 (inorganic, organic and drugs).
† Includes S.I.C. 284, 285, 287 and 289 (soap, paints, fertiliser and misc.).
‡ Includes in addition to fabricated metal products, ordnance, watches and clocks, optical equipment, and the S.I.C. 39 miscellaneous category.

Source: F. M. Scherer, Firm size, market structure, opportunity and the output of patented inventions. *American Economic Review*, 55 (1965) 1096–1125.

dustries appears to be suggested by these results.

Thirdly, Grabowski (1968) had earlier found that his multiple

TABLE 8.2 Inter-industry differences for the eight largest firms in the ratio of R & D professional employees to total sales*

Industry	*No. of firms*	*Mean (employees per million dollars of sales)*	*Coefficient of variation*	*Concentration ratio*†
Drugs	8	2.15	0.52	30.5
Industrial chemicals	6	1.51	0.33	55.1
Fertilisers	6	1.36	0.40	30.9
Paints and varnishes	6	1.27	0.25	36.4
Soaps and detergents	8	0.93	0.47	65.1
Metalwork machinery	4	0.81	0.27	17.0
Electric appliances	6	0.72	0.50	50.0
Perfumes and cosmetics	4	0.63	0.56	25.0
Gypsum, asbestos and misc. stone products	8	0.57	0.52	48.4
Motor vehicles	5	0.56	0.40	88.1
Metal cans	4	0.51	0.43	80.0
Petroleum refining	8	0.49	0.54	33.0
Flat glass and containers	7	0.45	0.28	68.9
Gen. industry machinery	7	0.42	0.41	35.7
Coating, engraving and misc. metal products	6	0.41	0.51	26.2
Distilled liquors	4	0.33	0.46	64.0
Canning and preserving	6	0.27	0.30	32.3
Sugar	4	0.27	0.45	65.0
Cotton goods	6	0.27	1.07	18.0
Construction and farm equipment	6	0.26	0.53	24.7
Confectionery	4	0.26	0.72	36.4
Grain mill products	6	0.25	0.70	32.7
Pulp, paper and products	8	0.23	0.82	23.8
Cigarettes	5	0.19	0.44	79.4
Smelting: copper, lead and zinc	6	0.18	0.63	72.2
Steel	8	0.13	0.38	56.0
Meat packing	7	0.13	0.48	32.7
Dairy products	6	0.10	0.73	37.6
Bakery products	5	0.09	0.70	30.5

* Mean and coefficients of variation (standard deviation ÷ mean) computed for R & D to sales ratios. Sales are for 1955. R & D employment figures are from National Research Council, *Industrial Research Laboratories of the United States*, 10th ed. (1956); sales data are from *Moody's* (1956).

† Concentration ratios are for 1954. See J. Stigler, *Capital and Rates of Return in Manufacturing Industries* (National Bureau of Economic Research, Princeton, 1963) p. 214.

Source: H. G. Grabowski and N. D. Baxter, Rivalry in industrial research and development: an empirical study. *Journal of Industrial Economics*, 21 (1973) 232, Table v.

regression analysis provided a good explanation of variation in research expenditures measured as a percentage of sales for the more research-intensive chemical and drugs industry in a pooled cross-section/time-series study of 16 chemical firms, 15 petroleum firms and 10 drug firms for the period 1959–62, drawn from the *Fortune* 500 listing for 1960 (see Table 8.3). Variation in research activity was analysed for each industry, with measures of current profitability, productivity of scientists, and diversification as independent variables.

TABLE 8.3. Estimation of regression equation
$$R_{i,t}/S_{i,t} = b_0 + b_1 P_1 + b_2 D_1 + b_3 (I_{i,t-1}/S_{i,t})$$
for the chemical, drug and petroleum industries for the period 1959–62

Industry	b_0	b_1	b_2	b_3	R^2	F	N
Chemicals	0.006	0.12*	0.019*	0.078*	0.63	29.76	60
	(0.004)	(0.02)	(0.004)	(0.023)			
Drugs	−0.03*	0.54*	0.41*	0.26*	0.86	73.71	40
	(0.01)	(0.12)	(0.07)	(0.05)			
Petroleum	0.002	0.016*	0.0049	0.020*	0.29	5.46	55
	(0.002)	(0.005)	(0.0071)	(0.00$)			

*Significant at the 0.01 level.

Note: Numbers below coefficient estimates are estimates of the standard errors; technological diversification variables (P_1 and D_1) have been multiplied by scale factors in order to present results more conveniently.

$R_{i,t}$ = level of R & D expenditures, *i*th firm, *t*th period;
$S_{i,t}$ = level of sales, *i*th firm, *t*th period;
$I_{i,t-1}$ = sum of after-tax profits plus depreciation and depletion expenses;
P_i = number of patents received per scientist and engineer employed by *i*th firm from 1955–9;
D_i = index of diversification of the *i*th firm (number of separate five-digit S.I.C. product classifications in which it produces).

Source: H. G. Grabowski, The determinants of industrial research and development: a study of the chemical drug and petroleum industries. *Journal of Political Economy*, (1968) p. 298.

However, Grabowski's model did not provide a good explanation of variation in research and development expenditures in the petroleum industry. Grabowski explains this partly by structural factors in the petroleum industry, such as the orientation towards process rather than product R & D, and the low degree of diversification in the petroleum industry. Competition still plays a role, however: 'Furthermore where R & D is a competitive strategy of lesser importance, as in petroleum refining, allocations to it tend to be more vulnerable to fluctuations in other uses of scarce funds' (p. 299).

Here again, competitiveness is called in to explain variability in inventive activity. There appears to be a consensus in the three studies with respect to variability in innovative activity: competitiveness in R & D leads to predictable and stable allocations. The strong version of the competitive hypothesis (Scherer, 1965; Grabowski and Baxter, 1973) suggests that oligopolistic rivalry in technological competition forces matching behaviour, while the weaker version (Grabowski, 1968) suggests that in less progressive industries R & D is a peripheral activity which is not a stable feature of corporate strategy.

However, Grabowski's results directly contradict the inference drawn from the findings of the other two studies. Grabowski's analysis differs from the other two in that it attempts to explain *differences* in innovative activity in respective industries, whereas the competitive matching hypothesis purports to explain *similarities* in innovative activity in particular industries. If matching behaviour were adopted by firms within Grabowski's industries, then research intensity of each firm would gravitate to a uniform industry value. As Grabowski points out, 'imitation by a firm of an "industry" ratio would imply less variability in the dependent variable and cause the intercept of the regression equation to become statistically significant relative to the explanatory variables postulated above' (p. 298). However, as he also points out (p. 298), the estimates of the intercept terms in his equations tend to cast doubt on this proposition.

Yet the inconsistency with the conclusions of the other two studies is even stronger than is suggested by these relationships; examination of Grabowski's equations (Table 8.3) reveals the paradoxical situation that the influence of the independent variables actually appears to *increase* as the technological progressiveness of the industry increases. In Grabowski's own words:

> the size of the regression coefficient associated with each of these variables increases with the research orientation of the industry involved – being the lowest in the petroleum industry and the highest in the drug industry in every case. Thus as research looms more important as a competitive strategy to the firms of an industry, each of our independent variables exerts a correspondingly greater effect on the level of research that a firm performs. (p. 298)

A further feature of the regression analysis is brought out if the values of the variables are examined. Grabowski does not provide these, but they may be calculated from the coefficients and the estimates of standard errors in Table 8.3. They are summarised in Table 8.4, the figures in parentheses indicating the rank order by size of the t values for each b coefficient. It is clear from Table 8.4 that the order of the individual t values in general also parallels the order of research orientation of the industry, with a single reversal, chemicals having a higher t value than drugs for the b_1 coefficient. Thus, as well as the overall goodness of fit of the particular regression equation improving as research becomes a more important competitive strategy in the respective industries, the value of the variance of R & D expenditure explained by individual independent variables to the residual error also increases in the same direction.

TABLE 8.4

	t values				
Industry	$t(b_0)$	$t(b_1)$	$t(b_2)$	$t(b_3)$	N
Chemical	1.5 (2)	6 (1)	4.75 (2)	3.39 (2)	60
Drugs	3 (1)	4.5 (2)	5.85 (1)	5.20 (1)	40
Petrol	1 (3)	0.69 (3)	0.69 (3)	3.33 (3)	55

Source: Computed from Table 8.3.

It may be appropriate at this juncture to summarise the aspects discussed above of the findings of the three studies. The inference drawn by both Scherer (1965) and Grabowski and Baxter (1973) from the relationship between size-corrected innovative activity and variability in such activity in respective industries is that competitive matching results in decreased variability relative to size. On the other hand, the Grabowski (1968) analysis suggests that not only do independent variables influence corporate allocations within industries, but also that the magnitude and statistical significance of the effect of each variable tends to increase as the research orientation of the industry increases, contrary to the expectations of the competitive matching hypothesis.

Yet there is an interesting link between all three studies. *In each case, explained variability in innovative activity tends to increase with the technological progressiveness of each industry.* This is despite the fact that in the Scherer and Grabowski–Baxter studies, explained variability in innovative activity depends on firms gravitating towards a *common* industry solution in terms of intensity of innovative activity, while in the Grabowski analysis, explained variability in innovative activity depends on the ability of the regression equations to explain *differences* between firms' allocations in each industry. Grabowski's analysis is blatantly inconsistent with the ideas of competitive matching in technologically progressive industries despite the apparent support for it from the other two studies. There is therefore a curious similarity in terms of the relationship between explained variability of corporate innovative activity and technological progressiveness in all three studies, despite the fact that in Grabowski's analysis the result is dependent on *non*-imitative behaviour, while in the other two studies it is explained in terms of imitative matching of corporate innovative activity.

Grabowski does suggest that 'more subtle and complex forms of imitation are best analysed in a more disaggregative context than the data permit here' (1968, p. 297). However, he does not elaborate on the implications of this statement and in fact its practical relevance must be extremely limited. Quite simply, competitive matching can only account for similarities in corporate behaviour, while more powerful theories (such as neoclassical theory in the broadest sense) attempt to account for both similarities and differences. Imitative hypotheses alone cannot account for the good fit of Grabowski's equations in the more technologically progressive industries.

In the next section a reconciliation of the apparently conflicting evidence will be suggested using a simple learning hypothesis based on the analysis of earlier chapters.

RESOURCE ALLOCATION ADAPTATION IN THE SYSTEMS APPROACH

In Chapter 5 the equation used to describe the relationship between a subsystem's share of overall corporate resources and other variables was

$$V_i = f_2(\phi_1, \ldots, \phi_m, I_1, \ldots, I_n)$$

where the ϕ variables were intra-firm characteristics and the I variables were environmental or industry-level characteristics. In the same chapter, informational *feedback* was identified as the mechanism which built up the preferred system on which the allocation decision is based. Assuming the existence of pattern and regularity in relations between firm and environment, the preference system is developed through firm–environment interaction and learnt through feedback from past allocations.

In Chapter 6 the circumstances in which a stable preference system for resources could be built up were discussed, and compared with actual budgeting conventions. Provided management had extensive experience of R & D resource allocation, and also that component projects did not dominate the R & D programme, there would be opportunity for the build-up of a stable resource preference system assuming that there was regularity in firm–environment relations. As expected, stability in resource budgeting tended to be associated with those areas in which these conditions had an opportunity to operate.

Chapter 6 was merely concerned with *attainment* of a stable resource preference system; it was not intended to discuss problems of 'appropriateness' or 'aptness' of the particular preference system and derived allocations. Yet inferences as to such considerations are suggested by the nature of the feedback facilitating build-up of resource preferences, and it is this aspect of the systems model which is developed below in attempting to account for the differences in findings of the three studies.

We would not expect the management of all firms to be equally expert at identifying the appropriate steady state of allocations for their own firms, any more than we would expect the management of all firms to have an opportunity to express their preference system in resource terms. Instead, expertise and ability to estimate the appropriate steady-state values of R & D allocations are liable to depend on the accumulated knowledge of allocations in this area processed by the relevant corporate management. *Since the feedback process provides a continuing source of relevant information, we would expect effectiveness of steady-state allocative decisions to depend on the accumulated experience of management in the relevant decision areas.* Effectiveness can be interpreted as being inversely related to the relative importance of the error component in decision-making.

This is a modest hypothesis which is consistent with observed behaviour in many areas involving learning. To continue the analogy of the billiards player discussed in the section on feedback in Chapter 5, we would expect performance to improve with practice, if we measure performance in terms of degree of error associated with shots of comparable difficulty. Similarly, as far as organisational performance is concerned, Starbuck (1965) states that 'young organisations have little experience in distinguishing important problems from unimportant ones' (p. 61) and 'Young firms experiment and misallocate more than old firms. This produces a variance component which decreases as the firm grows older' (p. 119).

Practice in the case of the billiards player, and age in the case of organisations, may therefore be regarded as proxies for accumulated experience. We shall re-examine the findings of each of the three studies in the light of this interpretation of the resource feedback process.

ADAPTIVE LEARNING IN CORPORATE ALLOCATIONS

It is appropriate that we look at the Grabowski analysis first, since it provides the direct challenge to the 'competitive matching' interpretation of corporate behaviour. Is there a simple explanation of the Grabowski findings suggested by the discussion of the previous section? In particular, can we explain why both the statistical significance of the regression equation as a whole, and of the effect of individual variables, tends to increase with the research orientation of the respective industries?

This may be achieved if we interpret research orientation in Grabowski's analysis as proxy for experience. The more technologically progressive and R & D intensive the firms in a particular industry, the greater their experience of R & D allocations and knowledge of potential utility. Thus, the error component in corporate allocations is a function both of time and level of activity in the appropriate area, since accumulated experience is interpretable as a direct function of both latter variables.

Therefore, as the general level of experience of firms increases with the technological progressiveness of the industry in which they operate, we would *expect* a corresponding improvement in the goodness of fit of industry regression equations as well as a general tendency for the ratio of the variance of R & D expenditure explained by individual independent variables to the residual error (t values) to increase in the same direction. This is consistent with a process of adaptive learning occurring as experience of R & D allocations accumulates; the appropriate steady state with respect to a particular ϕ variable will become more distinct to corporate management as experience increases, *ceteris paribus*, and will therefore tend to be clearer to firms operating in the more technologically progressive industries. Therefore, the general behaviour of t values of R^2 in Grabowski's analysis is consistent with a process of adaptive learning through negative feedback.

It would be useful if other studies of determinants of R & D *by industry* were available. In fact there is a dearth of such analysis and, as far as is known, only Scherer in his 1965 article (partly discussed earlier) and Howe and McFetridge (1976)[5] provide analyses comparable to that undertaken here by Grabowski. Scherer divided the firms in his sample (see Table 8.1) into four groups according to size of regression coefficient in Table 8.1, this being interpreted as a measure of 'technological opportunity'. The groups were, in order of degree of imputed technological opportunity, electrical, chemical, 'moderates' and 'unprogressives'. The variable measuring innovative activity, P_i (patenting output), was estimated as a function of S_i (sales) with squared and cubic S_i terms included in the equation (i taking values 1 to 4 for each of the groups). The inferred relationship was therefore a non-linear one, and the R^2 for each industry grouping was 0.94, 0.74, 0.77 and 0.55 respectively in order of imputed technological opportunity. Thus, as with the Grabowski analysis, the rough trend in this case is for goodness of fit of the industry equations to increase with the technological progressiveness of the industry. The middle two industry groupings reverse this trend, but it should be noted their R^2 values are quite close to one another. Since the regression equations are non-linear, the simple competitive matching hypothesis is not adequate, in this part of Scherer's analysis, to explain the apparent relationship between R^2 values and 'technological opportunity' as measured by the regression coefficients.

Howe and McFetridge (1976) conducted an econometric investigation of determinants of levels of R & D expenditure in 81 Canadian firms in the electrical, chemical and machinery industries using pooled annual cross-section data over the period 1967–71. Independent variables were sales (again three variables were created with the addition of squared and cubic forms', after-tax profit, depreciation and government incentive grants, all measured for the ith firm in the year of sampling. The Herfindahl index of the particular industry was also used as independent variable.

One difficulty in comparing the Howe and McFetridge analysis with that of Grabowski is that the same industry may differ between countries (in this case, the United States and Canada) as far as technological progressiveness of the industry is concerned. However, a recent O.E.C.D. publication (1971) provides data from 12 member countries which indicate that although an industrial branch's share of R & D expenditure performed in a particular country may vary widely from country to country, an industry's *ranking* according to this measure tends to be similar in different countries (see table, p. 122). In particular, the report comments on 'the predominance of the electrically and chemically based industries in all the advanced Member countries; these two industries are always amongst the first three in national totals, with the exception of chemically based industries in Sweden' (p. 121). As far as Canada is concerned, in 1963 these industry groupings had the highest share of R & D expenditure in the country with 24.6 per cent and 16.1 per cent respectively. Even though machinery is aggregated with metal products, it constituted only 6.3 per cent of the country's R & D. Given this international pattern of distribution of R & D activity, it seems that electrical and chemical industries are usually among the most technologically progressive measured in terms of general level of sectoral R & D activity. Canada is typical in this respect.

The Howe and McFetridge regression equations had recorded $\bar{R}^2$ values of 0.78 (electrical), 0.80 (chemical) and 0.27 (machinery). Thus there is a distinct difference in the goodness of fit between the two technologically progressive industries (electrical and chemical) and the machinery industry. As in the Grabowski analysis, the two technologically progressive industries have $\bar{R}^2$ values substantially higher than the third industry.

Therefore, taking the evidence of the Grabowski, Scherer, and Howe and McFetridge studies together, it would seem that there is a general tendency for goodness of fit to improve with technological progressiveness and, by inference, the opportunities for learning faced by the typical firm in respective industries. Before we go on to look at the other two studies mentioned earlier in the light of this interpretation, it may be useful to examine and compare the interpretation of the source of the error term in the adaptive interpretation particularly from Grabowski's point of view. Grabowski emphasises that the poor fit of the petroleum regression equation may be attributed to 'structural factors', such as process orientation of the industry, its degree of vertical integration and limited diversification, and the vulnerability of R & D to fluctuations in other uses of scarce funds. The high degree of unexplained variance in this equation is therefore attributed to specification error in the equation; Grabowski's explanation is that the equation is inappropriately specified in terms of the structural characteristics of the industry itself.

In the alternative explanation presented here, error is primarily due to decision-making error by the corporate management themselves.[6] It is implicitly assumed that the model-builder has provided a good specification of the behavioural factors influencing corporate allocations to R & D in this industry, and that residual error is a consequence of corporate inexperience and ignorance in estimating the appropriate steady-state allocations. This has important implications for model-building since it suggests that there is an unavoidable stochastic element in the regression equations which diminishes with the technological progressiveness of the industry. In this explanation, it is not the skill of model-builder but the skill of corporate management which accounts for differences in the goodness of fit of the regression equations.

However, while the adaptive learning interpretation is consistent with Grabowski's analysis, it is not immediately obvious how it may be reconciled with the findings of the other two studies. In the adaptive learning explanation we have in mind a functional relationship of the type described earlier, i.e.

$$V_i = f_2(\phi_1, \ldots, \phi_m, I_1, \ldots, I_m)$$

Yet the competitive matching hypothesis suggests that differences in the ϕ variables do not result in differences in allocations to innovative activity within a particular industry. The tendency for variability in innovative activity to diminish relative to size of firm and level of innovative activity in the Scherer and Grabowski–Baxter studies respectively, appears more obviously consistent with competitive matching than with adaptive learning in situations where ϕ variables are thought to be important.

A possible reconciliation may be achieved by considering possible differences in the relative strength of effect of the ϕ and I variables. If, for the range over which the respective ϕ and I variables, variation in I variables has a substantially greater effect on allocations to innovative activity compared to the ϕ variable, then *intra*-industry variation in innovative activity will be much lessthan *inter*-industry variation. In such circumstances, firms may appear to gravitate towards a common solution, not through direct imitation, but because internal and external determinants of steady-state solutions for firms *within* an industry lead to similar steady-state solutions.

From our previous argument there are two intra-industry sources of variability in innovative activity: variability due to ϕ variation, and random error due to imprecise knowledge of the appropriate steady-state value on the part of corporation management. We would expect the latter source of error to become relatively less important the more technologically progressive and research-oriented the industry. We would also expect that, as far as total explained variability in innovative activity is concerned, variability explainable in terms of I variables will tend to swamp the effect of the ϕ variables. These two effects together would be consistent with an apparent gravitation to a common industry solution as technological progressiveness of respective industries increases. However, such movement is illusory (as the Grabowski findings suggest) and is interpretable as quasi-imitative behaviour.

It is likely in the case of both the Scherer and Grabowski–Baxter studies that the actual range of variation in the ϕ variable in respective industries is not reflected in the samples themselves because of restrictions in both cases on the firms sampled. In the former the sample was selected from the 500 largest corporations for 1955, while in the latter the sample was restricted to the eight largest firms in each industry. Therefore the samples in both cases consisted of the top few corporations in each industry. The ϕ variable representing size will therefore operate over a restricted band in each industry, and it is probable that other potentially relevant ϕ variables such as degree of diversification, patent productivity, etc., will also operate over a limited range because of these sampling restrictions. To the extent that ϕ variables operate over a narrow band for each industry in the respective samples, their ability to create variability in innovative activity within a given industry is limited. In technologically progressive industries with ample opportunities for learning and adaptation to the appropriate steady-state level, this may be reflected in quasi-imitative behaviour.

An example of such quasi-imitative behaviour is shown in Fig. 8.1. Various measures could be used as a measure of innovative activity; in this case *R & D employment/sales* is chosen. Industry A is more research-oriented than industry B, with a mean level of sales-corrected employment of R & D personnel of E'' compared to E' in industry B. It is assumed that a linear formulation with intercept zero would provide the best specification of the relationship between level of innovative activity and size (measured in terms of sales) in each industry, and that the only ϕ variable affecting the intensity of innovative activity is size.

In Fig. 8.1, the effect of the Ø variable (size) is partially inhibited by the restricted size range over which the samples are taken. The size of firm in both industries varies between *Om* and *On*, and the least-squares regression lines are *Op* for industry A, and *Oq* for industry B. A characteristic of Fig. 8.1 is that while the corporate allocations in the research-intensive industry are grouped more closely around the respective lines based on the assumed linear specification of the industry relationship, they also cluster more closely around the mean E'' of industry A since there is a reduction in error due to the learning effect. Thus, an intra-industry investigation of the relationship between size and research intensity would report a higher R^2 and t value in industry A compared to B (as does Grabowski's analysis), while a crude comparison of variability in research intensity relative to mean research intensity would tend to provide evidence of gravitation towards a common

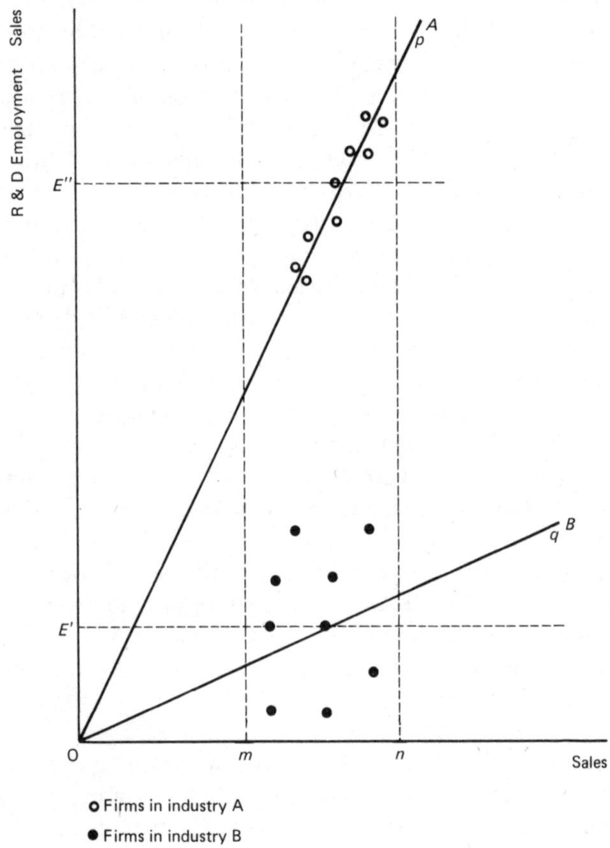

FIG. 8.1

industry solution (as do the Scherer and Grabowski–Baxter analyses). Both types of behaviour can be demonstrated in this example, but in fact this is a consequence of quasi-imitative adaptive learning and substantial suppression of the effect of variability in the ϕ variables.

Thus, in cases where variability in innovative activity created by ϕ variables is relatively unimportant compared to variability created by I variables, firms may appear to gravitate to a common industry solution even if the learning process in each case operates relatively independently. Note that it is not *necessary* for variability in ϕ variables to be low for these conditions to hold; corporate allocations to innovative activity may be relatively insensitive to variation in the ϕ variable compared to *I* variables, even in the face of high variability of ϕ variables. However, using samples of firms selected by size from particular industries is liable to reduce the potential effect of variation in ϕ variables on corporate allocations to innovative activity within a particular industry.

Therefore, the adaptive learning hypothesis offers a tentative explanation of the quasi-imitative behaviour identified in the Scherer and Grabowski–Baxter studies as well as the behaviour of the regression equations in the other, apparently conflicting studies.[7] This interpretation appears to offer a good general explanation of the characteristics of variability in innovative activity both across and within industry.

CONCLUSIONS

The above analysis suggests that rivalry and imitative behaviour is neither necessary nor sufficient to provide an adequate explanation of variation in innovative behaviour between firms. Instead, an adaptive learning hypothesis is tentatively suggested as an alternative interpretation in this area; provisionally, and with qualifications, it appears to provide a reasonable explanation of variation in innovative behaviour. This does *not* mean that the concept of rivalry or imitation is redundant in analysis of innovative behaviour; it may prove useful in analysing the process of adjustment to the steady state, especially at a more disaggregated level than is achieved here. The main contention in this analysis is simply that it is not necessary to invoke the concept of competitive matching in order to arrive at a simple explanation of variation in innovative activity. More importantly, simple competitive matching is not sufficient to account for intra-industry variability in innovative activity except as an error component, whereas it appears that adaptive learning may be adequate in this respect.

A crucial difference between the competitive matching and adaptive learning hypotheses lies in the interpretation of uncertainty. The rivalry studies and models tend to emphasise the high degree of uncertainty under which R & D decisions are made (even if subsequently they adopt a deterministic model). Decision-making in technologically progressive industries is consequently a highly uncertain occupation. The adaptive learning hypothesis on the other hand suggests the very opposite – as far as the R & D budget itself is concerned. In technologically progressive industries, opportunities for learning result in a lower degree of uncertainty as to what the appropriate steady-state level should be. The ability to estimate the appropriate level of R & D increases with practice and experience.

This suggests that the process of adaptation might be best observed over time, rather than by inference from cross-sectional or mixed cross-sectional time-series studies such as those above. This is indeed a possibility for future analysis, but there are two comments worth emphasising in this respect. Firstly, as far as the argument in this chapter is concerned, we were concerned with finding a simple explanation of apparently conflicting evidence, not with the mechanics or process of adaptation itself, and for this the adaptive learning hypothesis sufficed. Secondly, in a time-series analysis based on a period of time long enough to observe the process of adaptation reliably, it may be that the relationship between independent variables and requisite steady-state allocations may not remain constant: substantial environmental changes may shift the appropriate specification of the equations and the values of regression coefficients, and consequently different steady-state allocations may be appropriate at different points in the adjustment process. Therefore, identifying adaptation through learning over time may be more complex a problem than at first sight.

However, it may be the case that the nature of adaptation in a particular industry implies a reactive rivalry model. Environmental changes may be exogenous as far as a particular industry is concerned, e.g. changes in corporate tax may have an across-the-board effect on expected utility of specific allocations; but they may also be *endogenous* for a particular industry: changing market and technological characteristics are liable to be a consequence of corporate development and adjustment, with corporate action and reaction helping to create the future environment for the group of firms in a particular industry. In such circumstances a reactive learning model may indeed be useful for analysing the dynamic process of adjustment. Further scope for reactive models may exist for the special cases in which both ϕ and I variables are similar for a subset of firms within an industry; in this set of circumstances, firms may take advantage of indirect or 'vicarious' learning by imitating the allocations of the most successful firms.

As far as its relationship to the rest of the book is concerned, this chapter may be regarded as a development of Chapter 6, with its emphasis on perception and 'Gestalt' at high levels of abstraction in the large modern corporation. Separability of budget and project decision-making also plays a central role in the argument of both pieces of analysis.

With respect to the analysis of the previous chapters, the argument developed here may appear to conflict with the 'representative firm' assumption of Chapter 7, since ϕ variables are shown to have a significant effect on corporate allocations in research-intensive industries. However, in another sense it may be regarded as *supporting* the representative firm assumption since the analysis here suggests that I variables play a much more important role in creating variability in allocations to innovative activity than do the ϕ variables when the size range is limited. In Chapter 7 firms were grouped by industry and by two size bands, 1000–4999 and over 5000. Even within this basically industry-level analysis, however, some allowance is made for the effect of size of firm by including Z_{ij} in the analysis. Therefore, while Chapter 7 is obviously a fairly crude industry-level analysis, this chapter tends to reinforce the claim that even at this high level the analysis is capable of picking out some important determinants of innovative activity in the firm.

In summary, it has been suggested in this chapter that a simple adaptive learning hypothesis may account for the apparently inconsistent evidence of the studies cited. Competitive matching through rivalry is not required as a behavioural mechanism, and indeed it is extremely limited in terms of its ability to explain observed behaviour. The analysis here may be regarded as an application of the concept of corporate management as pattern-forming steady-state seekers. While the analysis may be interpreted as an empirical study in its own right, it has broader implications beyond the scope of the three studies cited. Considered with Chapters 6 and 7 it may be regarded as contributing to the idea of the corporation as a holistic, hierarchically organised system in which behaviour at higher levels may not be exclusively studied as aggregative phenomena, whether implicitly or explicitly assumed.

This last point has been discussed at length in previous chapters, and in the final chapter we shall attempt to define its importance for economic model-building more precisely. The argument is a simple one, but, as we shall see, it has generated substantial criticism and scorn from some eminent neoclassical economists. It is tentatively hoped that the usefulness of such an approach has been demonstrated in the last three chapters, and that the final chapter may help place it in an appropriate perspective.

9 Technological Change in the Modern Corporation and Implications for the Theory of the Firm

As was suggested at the beginning of this analysis, its purpose may be regarded as twofold. Its first and most obvious objective was the study and analysis of the factors affecting the allocation of resources to research and development activity in the large modern corporation, this being the concern of Chapter 7. The regression analysis of this chapter was conducted in an attempt to establish possible determinants of R & D activity; the results were consistent with a number of the hypothesised relationships. In particular, two results are worth emphasising for their implications for policy purposes. Firstly, no evidence was found to support the conventional wisdom that federal funding of R & D merely substitutes company funding; rather the regression analysis was consistent with the hypothesis that federal funds augment rather than replace company funds for R & D, *ceteris paribus*. Secondly, contrary to the initial expectation that technological opportunity (P_j) and propensity to undertake basic research (N_{ij}) are positively related, a statistically significant relationship between P_j and N_{ij} was found suggesting the opposite relationship (after the effects of industrial growth and size of research establishment had been separated out).

The implication for federal funding of R & D suggested by the first finding above is that there is no evidence that the opportunity cost of federally financed R & D includes corporate R & D that would have been undertaken in the absence of federal funding; this has direct relevance to any attempt to assess the economic effects or consequences of federal funding of R & D. As far as the second result concerning the possible relationship between N_{ij} and P_j is concerned, this at first appeared to constitute a puzzling contradiction to earlier expectations based on the spillover model. In fact the result prompted a reconsideration of the probable role of the P_j variable within this framework, and helped to indicate why technologically progressive firms might be more unwilling to undertake basic research than firms operating in relatively unprogressive industries. As a consequence, the special nature of the P_j variable with respect to other independent variables was emphasised and a counter-intuitive and apparently paradoxical result accounted for.

Yet the manner in which this reconciliation was achieved has a direct bearing on the second main objective of the analysis, the provision of a satisfactory framework for the analysis of resource allocation in the large modern corporation. The identification of possible subsystems within the R & D function facilitates the analysis of the possible effects of P_j. More generally, as far as the hypothesis formulation of the rest of the chapter was concerned, the resource-based systems approach provided fewer theoretical difficulties compared to traditional project-based approaches; in particular, it circumvented the problems of uncertainty and non-replicability of R & D projects by reorienting analysis from projects to resources. Perhaps even more significantly, the derived model is entirely consistent with the widespread convention of hierarchical 'top-down' resource allocation in the modern corporation as well as the tendency for management to base resource allocation on a fairly stable set of resource preferences.

The applicability of the systems approach is best illustrated by considering Chapters 6, 7 and 8 together. As well as providing what appears to be a more satisfactory framework for analysis in Chapter 7, the systems perspective suggests how differing conventions for R & D budgeting (described in Chapter 6) may be attributed to different circumstances for particular groups of firms. Chapter 8 contributed to the overall analysis by suggesting how the apparently conflicting evidence of different sets of studies could be reconciled by using an adaptive learning interpretation based on the systems approach of Chapter 5.

The thread common to these three chapters is the belief that by applying the concept of resource utility within a general systems framework, a simple description of corporate decision-making can be developed and satisfactorily applied to areas which project-based models find difficult or impossible to deal with.

However, it is *not* suggested that the systems approach as formulated here should be regarded as preferable to all project-based approaches whatever the circumstances. Theory based on satisficing behaviour may be applicable to certain types of problems in particular circumstances, and so also project models may be applicable to certain kinds of lower-level intra-functional or intra-divisional allocative decisions. A more general model might attempt to integrate these approaches; as far as behavioural theory is concerned, its assimilation within a more general open system approach would be obvious and natural given the similarity of satisficing to a homeostatic mechanism, while a possible project-based approach which may be promising in this respect is discussed later in this chapter.

For this last possibility alone, it would be appropriate to clarify how a resource-based systems approach such as that developed in this thesis might relate to project-based approaches developed in the same area. However, there is an even more urgent reason why such an examination should be conducted, and that is because the resource model counters one of the fundamental tools of traditional economic analysis – the technique of aggregation. This has extremely important implications for the interpretation of the resource utility concept which is not derivable from any examination or manipulation of individual product or project characteristics alone, and it is not reducible or divisible to such elements. It is appropriate to analyse closely the legitimacy of such procedure, since, as we shall see, previous suggestions that aggregation is not the sole means of analysing higher-level economic phenomena have met with strong resistance in some cases. To do so, it will be useful to consider first of all the concept of *emergence*.

'Emergence' is used by Medawar (1974) in discussing the analysis of complex systems when he describes the 'emergence at each tier of the hierarchy of concepts peculiar to and distinctive of that tier, and not obviously reducible to the notions of the level immediately above or higher still' (p. 57).

The relationship of emergence and reducibility is defined further by Becker (1974):

> A common philosophical strategy is to define emergence in terms of 'reducibility'. In the special case of hierarchically organised systems, an orthodox definition, neglecting refinements, would be something like this; *i*-level phenomena are 'emergent' with respect to lower level theories when, and only when, the *i*-level theories are not reducible to the theories of the lower levels. (p. 166)

The concept of resource may be regarded as demonstrably emergent at the higher levels of decision-making in the modern corporation. It is this argument in an economic context which was described earlier as being important contributions of Cyert and March and Edith Penrose. The latter chose to define resources independently of the concept of factor of production with its project-specific connotations (1959, p. 25). In neoclassical analysis the only relevance a factor has is in terms of its particular contribution or role in specific projects or operations. Analysing resources without defining them in terms of project-specific factors of production makes no sense in this approach.

In this respect, neoclassical analysis may be interpreted as a reductionist theory of the firm. According to Thorpe (1974), 'Reductionism is the attributing of reality exclusively to the smallest consitutents of the world, and the tendency to interpret higher levels of organisation in terms of lower levels' (p. 110; after Barbour, 1966).

In neoclassical analysis the smallest constituents of the economic world are the consumer and the individual project or product.[1] Macro-level analysis is conducted in terms of these concepts, and aggregation of micro-phenomena is the means employed to describe higher-level phenomena. This reductionist perspective may be applicable in some contexts, but its ability to contribute to the understanding of the process whereby corporate resources are allocated to technological change is extremely restricted. On the other hand, specifying composite resource as a concept emergent at higher levels of organisation does appear to have facilitated the analysis of the resource allocation in systems where higher-level allocation of resources generally precedes allocation to projects.

The differences of interpretation of the neoclassical theory and the open system resources model demonstrate what Weiss (1969) describes as:

> . . . the fundamental distinction between atomistic, micro-mechanistic terms of explanation on the one hand, and hierarchical concepts of organisation on the other. The difference is that the latter imply some sort of discontinuity encountered as one crosses interfaces between lower and higher orders of magnitude, while the former, trying to reduce all phenomena to the properties of ultimate elements in their various complications, are based on the premise of a continuity of gradations all the way up from the simple elements to infinite numbers of them. (pp. 8–9)

As Medawar (1974, p. 62) points out, this discontinuity of conceptualisation is apparent when the empirical sciences are arranged in hierarchical orderings. He selects five in the ordering (1) physics, (2) chemistry, (3) biology, (4) ecology/sociology, and comments that when the sciences are considered in that order, the degree of empirical content increases progressively, and new concepts emerge at each level which did not appear in the preceding science. When we come to level (4) in Medawar's scheme, theories and concepts specific to the social sciences begin to emerge which are neither apparent nor applicable in the frame of reference of the preceding sciences; to use Medawar's extreme example, 'a contextually distinctive notion like the "foreign exchange deficit" cannot be envisaged in the world of physics' (p. 62).

Emergence of concepts is observable within disciplines as well as between them. Psychology has a number of subdivisions, but two broad areas of concern are physiological psychology and social psychology. In the latter, concepts such as 'role' and 'coalition' are emergent with respect to physiological psychology, and concepts used in physiological psychology such as 'synaptic lapses' and 'central nervous system' are redundant in analysis of most social situations. *The emergent concepts displace the lower-level concepts as a general basis for analysis.*

The burden of the preceding chapters has been that emergent concepts at higher levels of economic behaviour must be recognised and developed if economics is to account for social organisation of resource allocation. This is easier said than accomplished, since attempts to do so or to suggest that this is either necessary or possible have frequently been met with rejection and even ridicule from neoclassical/reductionist theorists. A good example is the response to Vining's comment (1949a) on an article by Koopmans. The relevant criticism by Vining is quoted in Koopmans' reply (1949) below:

> I cannot understand the meaning of the phrase that 'the aggregate has an existence apart from its constituent particles and behaviour characteristics of its own not deducible from the behaviour characteristics of the particles'. If a theory formulates precisely (although possibly in probability terms) the determination of the choices and actions of each individual in a group or population, in response to the choices and actions of other individuals or the consequences thereof (such as prices, quantities, state of expectation) then the set of these individual behaviour characteristics is *logically*[2] equivalent to the behaviour characteristics of the group. (pp. 86–7)

Koopmans' view is strongly supported by Arrow (1968). With respect to the same comment by Vining quoted above by Koopmans, Arrow states:

> Taken literally this position seems indefensible. As Koopmans points out, a full characterisation of each individual's behaviour logically implies a knowledge of group behaviour; there is nothing left out. The rejection of the organism approach to social problems has been a fairly complete, and to my mind salutary, rejection of mysticism. (p. 641)

The justification for this complete condemnation of Vining's point of view is that, according to Arrow:

> In order to have a useful theory of relations among aggregates, it is necessary that they be defined in a manner derived from the theory of individual behaviour. In other words, even the definition of such magnitudes as national income cannot be undertaken without a previous theoretical understanding of the underlying individual phenomena. (p. 642)

Vining's choice of phrase is unfortunate since it parallels the naïve definition of 'Gestalt', 'the whole is greater than the sum of the parts', which was criticised and restated by Angyal (see Chapter 5). However, it is useful since it elicits a precise statement of the reductionist perspective of neoclassical theory by two eminent economists. There is no place for emergence in Koopmans' or Arrow's view of the world.

Yet what meaning has 'individual' in this view of the world? In Medawar's list of the empirical sciences, the consumer would be an *emergent* concept somewhere about the fourth level. The 'consumer' in this list would be an abstract concept not reducible to constituent atoms or molecules. The individual human being only appears at the beginning of the fourth level, and this level includes such higher-level organised systems as teams, institutions and countries. Looking at the numerous levels of emergent concepts included in the spectrum of the empirical sciences, the choice of 'the individual' as the exclusive behavioural concept applicable to economic activity appears to be supremely arbitrary. As Vining points out in his rejoinder (1949b), 'is it the individual that Koopmans regards as his unit anyway? Perhaps his unit is the family or the firm, in some instances a grouping of families and in many instances a grouping of firms' (p. 92).

The individual is a holistic concept no less and no more than the concept of the corporation developed in the preceding chapters. It is therefore as vulnerable to criticisms of 'mysticism' and logical redundance as the higher-level concept implicit in Vining's comment. However, hopefully the use of taxonomies such as Medawar's indicates both the relevance and limitations of such holistic concepts in their specificity to particular bands in the spectrum of the empirical sciences, and provides a perspective within which these concepts may be accommodated.

In this respect, the levels over which the resource-based model is applicable are strictly limited. Its application is restricted to a single level or a few levels in the organisational hierarchy, and to steady-state behaviour in those levels. One aspect which is of obvious relevance is what happens to the allocation of resources and resource activity once the level is reached where allocation to specific projects takes place. It is not directly obvious from the resources model how this may occur, but Nelson's study of R & D decision-making in the Bell Telephone Laboratories (1962) may be of relevance:

> Given the nature of scientific research and an organisation where individual scientists had a wide degree of freedom, the allocation of the scientific staff among competing alternatives is likely to be accomplished by an evolutionary or natural selection process . . . uncertainty and learning are key aspects of research . . . an alert scientist working on a project which appears to be running into sharply diminishing returns has very strong incentives – his professional reputation, his scientific curiosity and his future at the laboratories – to phase out his current work and phase in research in a more promising area – a new project or a going project which has exciting prospects. (p. 572).

This 'evolutionary' or 'natural selection' process would fit well in a resource-based model, since like such a model it presupposes the independent existence of the resource, in this case the individual scientist. It is also based on an evolutionary or learning process as is the process of adaptation hypothesised at the level of the corporation in Chapter 8. Nelson, in collaboration with Sidney Winter, has been developing in recent years a 'natural selection' approach to technological change at micro- and macro-levels in industry. A current statement of the general approach is made in Nelson and Winter (1974).

Therefore, there appears to be potential for the development and extension of the model developed in Chapter 5. While it is not intended to rebut completely the claims or rights of reductionist-type approaches to application in this area, it is felt that the above analysis may demonstrate the need to question the legitimacy of the view taken by Arrow and Koopmans above, that reductionism has the exclusive prerogative of investigation in economic behaviour. As long as institutions and organisations are regarded as simple aggregates by such theorists, the danger exists that a tremendous number of potentially rich theoretical models may be ignored or remain undeveloped. The relevance of the concept of emergence has been seen earlier in discriminating between physiological and social psychology, and indeed the attitude of die-hard neoclassical theorists may be compared to attempting to analyse the behaviour of individuals and groups in terms of physiological concepts; it may have limited feasibility, but certainly its general relevance is highly questionable.

In this respect, the general systems approach to technological change developed in this book may be regarded as a useful one. The observed behaviour of corporations in allocating resources to technological change has accorded well with the expectations of the derivative hypotheses, and aberrations were frequently accounted for in terms of the violation of basic behavioural assumptions of the theory. It is, of course, only one interpretation and application of general systems theory to economic behaviour, but in terms of its ability to describe the process of allocating resources to research and development in industry and to account for differences in R & D behaviour, it performs better than previous neoclassical theories. It is this which encourages optimism as to the possibilities of further application and development of general systems theory in the analysis of economic behaviour. While further investigation lies outside the scope of this analysis, obvious areas which may be highly amenable to investigation using a general systems framework are the marketing function within the firm (which, like R & D, has a high level of project uncertainty) and the public sector in general (which also tends to use a 'top-down' system of resource allocation).

However, the implications of accepting the legitimacy of systemic approaches in economic analysis are quite profound. It has been argued here that neoclassical theory is reductionist in philosophy, while the resources framework developed here is essentially systemic. The fundamentally incompatible nature of such approaches has been discussed in this chapter as well as in earlier chapters. Therefore, in developing a resources or systems framework which may be regarded as extending behavioural theory, we have sharpened and clarified the inherent conflict between the reductionist (neoclassical) view of the world and the systemic (extended behavioural) view of the world. The integration of behavioural theory within a general systems approach, which appears to have been both simple in theoretical terms and productive empirically, provides a 'top-down' frame of reference in what appears to be a 'top-down' world. Contrary to this, the neoclassical technique of aggregation provides a 'bottom-up' explanation of the same phenomena.

Therefore, a fundamental argument following from the analysis of this and previous chapters is that the economic theorist intending to apply theories explicitly or implicitly involving aggregation may be faced with a dilemma. This dilemma has been examined here in the context of resource allocation to R & D, but by implication it is liable to be a more general problem. He may have to decide whether to continue squeezing what appears to be a round world into the square hole of reductionist theory, or whether to accept that aggregation may be inferior to a non-reducible systemic approach in certain circumstances. If the latter decision is taken, then it constitutes a radical reappraisal of the way we look at economic behaviour.

APPENDICES

Appendix I

DATA MATRIX

Industry	Dependent variables					Independent variables						
	X_{ij}	Y_{ij}	B_{ij}	N_{ij}	P_j	G_j	U_j	E_{ij}	E'_{ij}	F_{ij}	V_j	Z_{ij}
1. Food	*	0.3	*	16	1	1.19	37	*	0.45	*	0.32	155
2. Textiles	0.6	0.7	*	*	①	1.24	33	0.24	0.28	0.85	0.43	41
3. Lumber	0.3	0.3	0	0	1	1.24	44	0.13	0.13	1.00	0.47	40
4. Paper	0.8	0.8	1.5	11	1	1.31	38	0.47	0.47	1.0	0.45	56
5. Chemicals	3.3	4.0	7.8	45	3	1.54	33	2.52	3.11	0.81	0.55	77
6. Indust.	*	5.5	*	44	③	1.63	32	*	5.06	*	0.55	92
7. Drugs	*	6.4	*	64	③	1.57	37	*	5.07	*	0.76	78
8. Other chem.	1.9	1.9	9.5	38	2	1.39	30	1.28	1.31	0.98	0.50	69
9. Ind./Drugs	5.1	5.9	7.2	53	③	1.62	34	3.83	5.07	0.77	0.58	86
10. Petrol	1.0	1.0	*	*	2	1.26	35	1.3	1.3	1.0	0.21	136
11. Rubber	1.0	1.0	*	*	①	1.55	33	0.42	0.43	1.0	0.51	44
12. Stone	*	1.1	7.1	31	①	1.25	33	*	0.54	*	0.57	51
13. Prec. met.	0.8	0.9	3.3	11	1	1.32	29	0.57	0.65	0.87	0.43	69

Industry	Dependent variables				Independent variables							
	X_{ij}	Y_{ij}	B_{ij}	N_{ij}	P_j	G_j	U_j	E_{ij}	E'_{ij}	F_{ij}	V_j	Z_{ij}
14. Ferrous	*	0.5	*	*	①	1.27	28	*	0.25	*	0.47	47
15. Non-fer.	*	1.3	*	*	①	1.44	29	*	1.33	*	0.35	104
16. Fab. met.	1.1	1.2	*	10	2	1.23	36	0.44	0.47	0.95	0.51	40
17. Machine	2.0	2.8	1.3	11	3	1.37	32	0.90	1.25	0.72	0.57	45
18. Other elec.	2.6	3.2	1.4	14	3	1.34	31	1.05	1.30	0.81	0.54	41
19. Motor	0.7	0.9	*	*	②	1.64	24	0.37	0.44	0.83	0.35	52
20. Aircraft	2.0	12.9	*	15	3	1.18	23	0.85	5.54	0.15	0.57	43
21. Instrum.	3.4	4.9	5.5	16	3	1.40	33	1.35	1.97	0.69	0.65	40
22. Sci.	*	3.1	*	*	③	1.20	34	*	1.19	*	0.66	38
23. Optic.	*	6.9	*	*	③	1.54	32	*	3.00	*	0.65	43
24. Other	0.8	1.4	3.3	17	①	1.21	30	0.33	0.56	0.6	0.57	40
25. Food	*	0.4	9.5	37	1	1.19	37	*	2.20	*	0.32	542
26. Textiles	0.4	0.4	7.1	*	①	1.24	33	1.0	1.0	1.0	0.43	259
27. Lumber	0.4	0.4	0	0	1	1.24	44	1.0	1.0	1.0	0.47	268
28. Paper	0.7	0.7	2.1	32	1	1.31	38	2.18	2.18	1.0	0.45	300

29. Chemicals	3.7	4.8	12.3	73	3	1.54	33	16.89	21.96	0.77	0.55	453
30. Indust.	4.4	5.5	13.0	83	③	1.63	32	23.96	30.13	0.80	0.55	546
31. Drugs	*	3.7	18.0	73	③	1.57	37	*	11.90	*	0.76	316
32. Other chem.	*	3.6	4.3	50	②	1.39	30	*	14.10	*	0.50	389
33. Petrol	0.9	1.0	15.9	*	2	1.26	35	12.23	13.14	0.93	0.21	1366
34. Rubber	*	2.6	6.3	*	①	1.55	33	*	18.14	*	0.51	711
35. Stone	*	2.1	7.0	67	①	1.25	33	*	6.73	*	0.57	316
36. Prec. met.	0.7	0.8	7.0	48	1	1.32	29	4.15	4.33	0.96	0.43	564
37. Ferrous	0.7	0.7	8.4	*	①	1.27	28	4.04	4.13	0.98	0.47	587
38. Non-fer.	0.9	0.9	4.2	*	①	1.44	29	4.40	4.80	0.92	0.35	513
39. Fab. met.	1.5	2.0	2.3	29	2	1.23	36	4.71	6.21	0.76	0.51	318
40. Machine	4.0	5.6	3.4	27	3	1.37	32	12.51	17.37	0.72	0.57	312
41. Other elec.	3.2	9.0	2.3	48	3	1.34	31	15.17	42.57	0.36	0.54	469
42. Motor	2.6	3.5	*	*	②	1.64	24	28.54	32.86	0.73	0.35	1101
43. Aircraft	2.6	27.0	1.2	56	3	1.18	23	1.63	172.93	0.01	0.57	641
44. Instrum.	5.5	10.7	*	67	3	1.40	33	15.0	29.0	0.52	0.65	272
45. Sci.	*	17.0	*	*	③	1.20	34	*	42.75	*	0.66	250
46. Optic.	*	7.8	*	*	③	1.54	32	*	22.13	*	0.65	283
47. Other	*	0.7	8.5	37	①	1.21	30	*	2.47	*	0.57	347

Note: The encircled values of P_j indicate values calculated by F. Swenson.

Appendix II

SAMPLES MATRIX

Industry	Sample					
	(*a*)	(*b*)	(*c*)†	(*d*)	(*e*)	(*f*)
1. Food						
2. Textiles				*		
3. Lumber	*	*	*	*		
4. Paper	*	*	*	*	*	*
5. Chemicals	*	*	*			
6. Indust.						
7. Drugs						
8. Other chem.				*	*	*
9. Ind./drugs				*	*	*
10. Petrol	*			*		
11. Rubber				*		
12. Stone						
13. Prec. met.	*	*	*	*	*	*
14. Ferrous						
15. Non-fer.						
16. Fab. met.	*		*	*		*
17. Machine	*	*	*	*	*	*
18. Other elec.				*	*	*
19. Motor				*		
20. Aircraft	*		*	*		*
21. Instrum.	*	*	*	*	*	*
22. Sci.						
23. Optic.						
24. Other				*	*	*
25. Food						
26. Textiles				*	*	
27. Lumber	*	*	*	*		
28. Paper	*	*	*	*	*	*
29. Chemicals	*	*	*	*	*	*
30. Indust.						
31. Drugs						
32. Other chem.						

Industry	*Sample*					
	(*a*)	(*b*)	(*c*)†	(*d*)	(*e*)	(*f*)
33. Petrol	*	*		*	*	
34. Rubber						
35. Stone						
36. Prec. met.	*	*	*			*
37. Ferrous				*	*	
38. Non-fer.				*	*	
39. Fab. met.	*	*	*	*	*	*
40. Machine	*	*	*	*	*	*
41. Other elec.				*	*	*
42. Motor				*		
43. Aircraft	*	*	*	*	*	*
44. Instrum.	*		*	*		*
45. Sci.						
46. Optic.						
47. Other						

† Sample (c) is the sample which would be used for N_{ij} excluding the additional P_j estimates, but is not used in the regression of Appendix III.

Appendix III

Regression equations for Chapter 7

Eq. no.	*Dependent variable*	*Independent variables*						*F*	R^2	$\bar{R}^2$	*Sample*
(1.1)	X_{ij}	-10.41	$+1.84P_j$	$+6.65G_j$	$+0.029U_j$	$+0.0043Z_{ij}$					
		-2.67*	4.52**	2.28*	0.48	1.66		14.20	0.81	0.76	(a)
(1.2)	B_{ij}	-24.18	$+7.81P_j$	$-12.48G_j$	$+0.19E'_{ij}$	$+62.67F_{ij}$	$0.70U_j$				
		-1.14	2.97*	-0.79	2.73*	3.44**	-2.55*	3.65	0.70	0.50	(b)
(1.3)	B_{ij}	5.39	$+1.13P_j$	$-0.97G_j$	$+0.52E_{ij}$	$+10.88F_{ij}$	$-0.37U_j$				
		0.28	0.44	-0.58	2.34*	1.02	-1.28	2.97	0.65	0.43	(b)
(2.1)	X_{ij}	-4.56	$+1.96P_j$	$+6.02G_j$	$-0.040F_{ij}$	$+0.0054Z_{ij}$	$-0.064U'_j$				
		-1.06	5.21**	2.91**	-0.24	2.60*	-1.08	18.78	0.81	0.77	(d)
(2.2)	$\log X_{ij}$	-5.71	$+1.16 \log P_j$	$+1.38 \log G_j$	$+0.096 \log F_{ij}$	$+0.14Z_{ij}$	$+1.23 \log U'_j$				
		-1.38	7.70**	1.73	0.96	2.42*	1.22	23.56	0.84	0.81	(d)
(2.3)	X_{ij}	14.71	$+\frac{5.45}{P_j}$	$+\frac{10.72}{G_j}$	$+\frac{0.0028}{F_{ij}}$	$+\frac{34.80}{Z_{ij}}$	$\frac{-88.30}{U'_j}$				
		4.24**	-6.53**	-2.86**	-0.19	-2.91**	0.40	17.66	0.80	0.76	(d)
(2.4)	$\log X_{ij}$	5.23	$\frac{+1.76}{P_j}$	$\frac{+2.24}{G_j}$	$\frac{+0.0020}{F_{ij}}$	$\frac{+-8.46}{Z_{ij}}$	$\frac{+77.94}{U'_j}$				
		5.71**	-7.97**	-2.27*	-0.52	-2.68*	-1.32	24.96	0.85	0.82	(d)

Eq. no.	Dependent variable	Independent variables						F	R^2	$\bar{R}^2$	Sample
(3.1)	Y_{ij}	31.23	$+1.06P_j$	$-3.02G_j$	$-28.39F_{ij}$	$+0.018Z_{ij}$	$-0.029U'_j$				
		1.97	0.76	-0.40	-4.77**	2.37*	-0.13	22.9	0.84	0.80	(d)
(3.2)	$\log Y_{ij}$	-9.49	$+1.36 \log P_j$	$+0.61 \log G_j$	$-0.38 \log F_{ij}$	$+0.12 \log Z_{ij}$	$+2.20 \log U'_j$				
		2.12	8.36**	0.70	3.48**	1.91	2.03	45.46	0.91	0.89	(d)
(3.3)	Y_{ij}	21.73	$+\frac{11.37}{P_j}$	$-\frac{10.99}{G_j}$	$-\frac{0.35}{F_{ij}}$	$+\frac{5.15}{Z_{ij}}$	$+\frac{1078.75}{U'_j}$				
		2.04	-4.45**	0.96	7.71**	-1.40	-1.58	33.10	0.88	0.86	(d)
(3.4)	$\log Y_{ij}$	6.17	$+\frac{2.35}{P_j}$	$-\frac{0.30}{G_j}$	$\frac{0.012}{F_{ij}}$	$+\frac{8.11}{Z_{ij}}$	$+\frac{226.09}{U'_j}$				
		4.99**	-7.90**	0.22	2.33*	-1.90	-2.84**	28.95	0.87	0.84	(d)

Eq. no.	Dependent variable	Independent variables						F	R^2	$\overline{R}^2$	Sample
(4.1)	B_{ij}	-29.83	$+1.53P_j$	$+3.75G_j$	$+0.049E'_{ij}$	$+16.91F_{ij}$	$+0.19U'_j$				
		-1.01	0.85	0.30	1.00	1.83	0.52				
(4.2)	B_{ij}	-33.30	$+0.73P_j$	$-1.57G_j$	$+0.43E_{ij}$	$+12.36F_{ij}$	$+0.40U'_j$	2.15	0.47	0.25	(e)
(4.3)	$\log B_{ij}$	-20.09	$-0.30 \log P_j$	$+1.44 \log G_j$	$+0.31 \log E_{ij}$	$+0.36 \log F_{ij}$	$+4.99 \log U'_j$				
		-1.12	-0.62	0.50	1.91	1.35	1.17	1.79	0.43	0.19	(e)
(4.4)	B_{ij}	13.98	$\frac{-1.43}{P_j}$	$\frac{+4.96}{G_j}$	$\frac{+2.72}{E'_{ij}}$	$\frac{+0.056}{F_{ij}}$	$\frac{+273.02}{U'_j}$				
		0.49	0.32	-0.21	-1.31	-0.84	-0.16	0.63	0.21	-0.12	(e)
(4.5)	B_{ij}	16.94	$\frac{-0.74}{P_j}$	$\frac{+2.40}{G_j}$	$\frac{+2.02}{E_{ij}}$	$\frac{+0.052}{F_{ij}}$	$\frac{+581.88}{U'_j}$				
		0.60	0.17	-0.10	-1.30	-0.79	-0.35	0.62	0.21	-0.12	(e)
(4.6)	$\log B_{ij}$	6.06	$\frac{-0.66}{P_j}$	$\frac{+2.17}{G_j}$	$\frac{+0.59}{E'_{ij}}$	$\frac{+0.016}{F_{ij}}$	$\frac{+207.40}{U'_j}$				
		1.20	0.83	-0.53	-1.61	-1.33	-0.70	1.26	0.34	0.07	(e)
(4.7)	$\log B_{ij}$	6.73	$\frac{-0.49}{P_j}$	$\frac{+1.70}{G_j}$	$\frac{+0.42}{E_{ij}}$	$\frac{+0.013}{F_{ij}}$	$\frac{+272.66}{U'_j}$				
		1.33	0.63	-0.40	-1.49	-1.25	-0.91	1.17	0.33	0.05	(e)

There are only 7 equations in the above run for B_{ij} since the doub. log. transform. for E'_{ij} was incorrectly specified with a wrong variable. From the general results indicated by the other regressions it was not thought worthwhile repeating the E'_{ij}/doub. log. transform. regression.

Eq. no.	*Dependent variable*	*Independent variables*						F	R^2	$\bar{R}^2$	*Sample*
(5.1)	N_{ij}	-104.14	$-7.75P_j$	$+149.00G_j$	$+0.28E'_{ij}$	$-2.86F_{ij}$	$-0.44U'_j$				
		-0.92	-1.12	3.20**	2.18*	-0.94	-0.34	3.88	0.62	0.46	(f)
(5.2)	N_{ij}	-109.39	$-9.85P_j$	$+103.73G_j$	$+2.01E_{ij}$	$-35.80F_{ij}$	$+0.60U'_j$				
		-1.03	-1.53	2.18	2.65*	-1.35	0.47	4.74	0.66	0.52	(f)
(5.3)	$\log N_{ij}$	-0.004	$-0.58\log P_j$	$+3.26\log G_j$	$+0.42\log E'_{ij}$	$+0.080\log F_{ij}$	$+0.52\log U'_j$				
		-0.00055	-2.59*	2.29*	5.38**	0.51	0.28	12.00	0.83	0.76	(f)
(5.4)	$\log N_{ij}$	0.28	$-0.58\log P_j$	$+3.36\log G_j$	$+4.18\log E_{ij}$	$-0.35\log F_{ij}$	$+0.45\log U_j$				
		0.035	-2.58*	2.35*	5.30**	-2.84*	0.24	11.72	0.83	0.76	(f)
(5.5)	N_{ij}	124.96	$\frac{-16.72}{P_j}$	$\frac{+160.03}{G_j}$	$\frac{+17.01}{E'_{ij}}$	$\frac{0.38}{F_{ij}}$	$\frac{-1830.80}{U'_j}$				
		1.22	1.12	-2.17	-2.71*	1.94	0.36	4.55	0.65	0.51	(f)
(5.6)	N_{ij}	163.19	$\frac{-13.66}{P_j}$	$\frac{+137.77}{G_j}$	$\frac{+13.27}{E_{ij}}$	$\frac{-0.36}{F_{ij}}$	$\frac{+1746.36}{U'_j}$				
		1.60	0.90	-1.72	-2.46*	1.80	-0.34	4.09	0.63	0.48	(f)
(5.7)	$\log N_{ij}$	5.08	$\frac{-0.73}{P_j}$	$\frac{+4.10}{G_j}$	$\frac{+0.73}{E'_{ij}}$	$\frac{-0.011}{F_{ij}}$	$\frac{-85.27}{U'_j}$				
		1.65	1.63	-1.85	-3.85**	1.84	0.56	6.18	0.72	0.60	(f)
(5.8)	$\log N_{ij}$	6.72	$\frac{-0.60}{P_j}$	$\frac{+3.17}{G_j}$	$\frac{+0.56}{E_{ij}}$	$\frac{-0.010}{F_{ij}}$	$\frac{+67.26}{U'_j}$				
		2.11	1.27	-1.27	-3.35**	1.61	-0.43	5.03	0.68	0.54	(f)

Eq. no.	Dependent variable	Independent variables				F	R^2	$\bar{R}^2$	Sample
(6.1)	X_{ij}	-8.06	$+1.87P_j$	$+5.55G_j$	$+0.0047Z_{ij}$				
		-3.69**	7.14**	3.36**	2.43*	30.94	0.79	0.77	(d)
(6.2)	$\log X_{ij}$	-0.69	$+1.14 \log P_j$	$+1.92 \log G_j$	$+0.13Z_{ij}$				
		2.36*	8.72**	3.00**	2.35*	39.36	0.83	0.81	(d)
(6.3)	X_{ij}	15.91	$+\frac{5.29}{P_j}$	$+\frac{10.73}{G_j}$	$+\frac{34.23}{Z_{ij}}$				
		6.81**	-7.04**	-3.66**	-3.04**	31.59	0.80	0.77	(d)
(6.4)	$\log X_{ij}$	4.40	$+\frac{1.82}{P_j}$	$+\frac{2.65}{G_j}$	$+\frac{8.14}{Z_{ij}}$				
		6.92**	-8.88**	-3.05**	-2.66*	41.42	0.84	0.82	(d)

Eq. no.	Dependent variable	Independent variables		F	R^2	$\bar{R}^2$	Sample
(7.1)	B_{ij}	−8.20	$+9.93G_j$				
		−0.68	1.10	1.21	0.07	0.01	(e)
(7.2)	B_{ij}	3.49	$+0.35E_{ij}$				
		2.88*	2.03	4.11	0.20	0.15	(e)
(7.3)	B_{ij}	5.14	$-0.014P_j$				
		2.04	−0.013	0.00016	0.0000099	−0.06	(e)
(7.4)	$\log B_{ij}$	0.34	$+3.37 \log G_j$				
		0.48	1.42	2.03	0.11	0.06	(e)
(7.5)	$\log B_{ij}$	1.08	$+0.28 \log E_{ij}$				
		4.85**	1.85	3.4	0.18	0.12	(e)
(7.6)	$\log B_{ij}$	1.37	$-0.087 \log P_j$				
		4.43**	−0.22	0.048	0.0030	−0.059	(e)
(7.7)	B_{ij}	18.33	$\frac{+17.61}{G_j}$				
		1.42	−1.03	1.06	0.06	0.0035	(e)

Eq. no.	*Dependent variable*	*Independent variables*		*F*	*R*	$\overline{R}^2$	*Sample*
(7.8)	B_{ij}	6.54	$+\frac{1.68}{E_{ij}}$				
		4.94**	−1.55	2.40	0.13	0.08	(*e*)
(7.9)	B_{ij}	5.83	$+\frac{1.16}{P_j}$				
		2.55*	−0.35	0.12	0.0077	−0.054	(e)
(7.10	$\log B_{ij}$	4.7	$+\frac{4.51}{G_j}$				
		1.92	−1.39	1.93	0.11	0.05	(e)
(7.11)	$\log B_{ij}$	1.58	$+\frac{0.34}{E_{ij}}$				
		6.08**	−1.44	2.08	0.12	0.06	(e)
(7.12)	$\log B_{ij}$	1.29	$-\frac{5.23}{P_j}$	0.0066	0.0004	−0.06	(e)
		2.89*	0.081				

Eq. no.	Dependent variable	Independent variables					F	R^2	$\overline{R}^2$	Sample
(8.1)	N_{ij}		-133.31	$-3.16P_j$	$+124.48G_j$	$+0.35E'_{ij}$				
			2.94*	-0.66	3.44**	3.41**	6.68	0.59	0.50	(f)
(8.2)	N_{ij}		-125.21	$+133.45G_j$	$+0.318E'_{ij}$					
			-2.93*	3.59**	3.56**		10.18	0.58	0.52	(f)
(8.3)	$\log N_{ij}$		2.09	$-0.60 \log P_j$	$+3.68 \log G_j$	$+0.40 \log E'_{ij}$				
			7.56**	-2.97*	3.81**	7.27**	22.73	0.83	0.79	(f)
(8.4)	$\log N_{ij}$		2.03	$+2.79 \log G_j$	$+0.30 \log E'_{ij}$					
			5.98**	2.47*	5.51**		19.55	0.72	0.69	(f)
(8.5)	N_{ij}		112.07	$-\frac{1.24}{P_j}$	$+\frac{98.26}{G_j}$	$+\frac{19.56}{E'_{ij}}$				
			2.39*	0.80	-1.54	-3.14**	5.57	0.54	0.45	(f)
(8.6)	N_{ij}	111.81	$+\frac{91.20}{G_j}$	$+\frac{16.94}{E'_{ij}}$						
		2.41*	-1.46	-3.23**			8.24	0.52	0.46	(f)
(8.7)	$\log N_{ij}$	5.35	$-\frac{0.62}{P_j}$	$+\frac{2.56}{G_j}$	$+\frac{0.78}{E'_{ij}}$					
		3.84**	1.36	-1.35	4.25**		8.34	0.64	0.56	(f)
(8.8)	$\log N_{ij}$	5.33	$+\frac{2.20}{G_j}$	$+\frac{0.65}{E'_{ij}}$						
		3.73**	-1.14	-4.04**			10.96	0.59	0.54	(f)

Notes

Chapter 2

1. This definition does not include quality control, routine product testing, market research, sales promotion, sales service, research in the social sciences or psychology, and other non-technological activities or technical services.

2. Schmookler (1962a, p. 43), however, classifies together basic and applied research activities as defined by the N.S.F. as applied scientific research. Schmookler justifies this by pointing out that industrial basic research may have eventual industrial application as its objective even if the nature of such application is difficult or impossible to specify. The objectives of basic research are difficult if not impossible to specify in economic terms *exante* and, as Nelson (1959a, pp. 300–1) points out, the loose or vague definition of goals at the basic research end of the R & D function is a rational response to the great uncertainties involved.

3. See, for example, Barnard (1948), Baumol (1959), Berle and Means (1932), Cyert and March (1963, chap. 2), Gordon (1961), Papandreou (1952) and Williamson (1964).

4. Nelson (1959a, p. 306) points out that this different orientation, and its implications for economic value to the information-producing unit, works back through the price system to create the differing salary scales for scientists and technologists.

5. Consequently, some taxonomies of R & D activity have attempted to identify scientific and technological subsystems of R & D activity, the former concerned with discovery of scientific knowledge, the latter concerned with revision and augmentation of productive techniques and products. They therefore emphasise output rather than objectives and differ from the N.S.F. definitions in this respect. See Machlup (1962a), Ames (1961), Kuznets (1962) and Schmookler (1962a, 1966).

6. Schon interprets innovation as commercial application of invention. However, since he identifies parallel rational views of invention and innovation, here we continue to use the broader interpretation of innovation used in the previous section.

7. Linear models typically predict that there will be a steady progression through basic research, applied research and development to production with *no backtracking*; shorter chains may be composed by eliminating earlier stages such as basic research. The linearity assumption can fit neatly into a rational model; if it is believed that a problem is routine and consequently programmable, backtracking would be redundant and inefficient. Backtracking only becomes legitimate managerial behaviour when the rational model is untenable, in particular when uncertainty is pervasive. Langrish et al. specifically identify

Blackett (1968) and Holloman (1965) as proponents of linear models of innovation. See also MacLaurin (1953) for development of a linear model.

8. 'Non-mission-oriented research (NMOR) is research carried on for the purpose of acquiring new knowledge, according to the conceptual structure of the subject or the interests of the scientist, without concern for a mission or application, even though the project in which such research is done may be funded with possible applications in mind' (Globe et al., 1973, p. 9).

9. 'Significant event is an occurrence judged to encapsulate an important activity in the history of an innovation or its further improvements, as reported in publications, presentations, or references to research' (Globe et al., 1973, p. 9).

10. Further specific criticism of the concept of a linear model of innovation is provided by Bryan (1973, pp. 31–3), Douds (1971, p. 74), Price and Bass (1969, pp. 802–3, 805), Marquis and Allen (1966, pp. 1053–4) and Encel (1970). Siegel (1962, pp. 447–8) also emphasises that technology can considerably affect the conduct of science in certain circumstances (e.g. in contrivance of various types of instrumentation). Bogaty (1969) points out the difficulties of advocating or operating the management and direction of research, and Freeman (1974, p. 40), Aram (1973) and Nelson (1972) cast doubt on the possibility of ever achieving rational, ordered and directed control over the innovation process. Aram's viewpoint is the most extreme, and touches on aspects of innovation to be dealt with in the next section: 'To a large extent . . . management and innovation are contradictory terms. So does . . . the concept of organising for innovation contain a contradiction: organisation strives to be a planned, controlled, predictable system of activities, while innovation is unplanned, uncontrolled and unpredictable' (p. 24).

11. Adjusted for subsequent variation in the price index and in the output required of the new development.

12. A recent example of disastrously miscalculated development cost and time is that of the Concorde. Government estimates in 1962 were that it would cost £150–170 million and enter service in 1967. By 1973 the Committee of Public Accounts revised estimate was £1065 million (unadjusted for inflation).

13. For further evidence see also Peck and Scherer (1962), Mansfield (1971, chap. 5, 'Overruns and Errors in Estimating Development Cost and Time', pp. 86–109) and Norris (1971).

14. The laboratory was Bell Telephone Laboratories, Nelson's analysis being primarily concerned with its invention of the point contact transistor (1948) and the junction transistor (1951), for which a Nobel prize was awarded in 1952 to three members of the development team (Nelson, 1962, p. 549).

15. Chapter V, 'Models for Change', pp. 112–38.

16. See also Klein and Meckling (1958) and Klein (1963).

17. Hirschman (1958) and Lindblom (1959) are used as examples of convergent thinking in economic development and public policy respectively.

18. We shall subsequently suggest that this is too narrow an interpretation of uncertainty for use in analysis of R & D.

19. In addition, see also for reference to the significance of uncertainty for innovative activity, Schmookler (1962a, p. 47), Sanders (1962, pp. 59–60), Klein (1963, pp. 478–84), Schon (1967, pp. 21, 24–32), Markham (1965, pp. 72–5), Jewkes et al. (1969, p. 105), Staff Report (I.R.I.) (1962), Shanks (1967, pp. 64–7), Hirschman (1967, pp. 75–81) and Carter and Williams (1958, chap. 3).

20. For examples of models incorporating potential redirection of R & D orientation by feedback from lower levels of R & D uncertainty to higher levels (e.g. development to applied research and applied research to basic research), see Jantsch (1969), Marquis and Allen (1966) and Gibson (1962).

21. See earlier references.

22. See Freeman and Bowie above, Gold (1971, p. 224), Langrish et al. (1972, p. 41), Nelson et al. (1967, pp. 85–6), Norris and Vaizey (1973, p. 66) and Carter and Williams (1957, p. 52).

23. In the context of the high uncertainty associated with the eventual commercial applicability of this subsystem it is possible to ignore the fact that a great deal of scientific endeavour can become fairly routine and predictable. Kuhn (1970) suggests that, in scientific inquiry in general, 'perhaps the most striking feature of the normal research problems . . . is how little they aim to produce major novelties, conceptual or phenomenal. Sometimes, as in a wavelength measurement, everything but the most esoteric detail of the result is known in advance, and the typical latitude of expectation is only somewhat wider' (p. 35). Normal science is research based on, and designed to articulate and develop, a generally accepted paradigm (Kuhn, 1970, pp. 23–4). However, while there would appear to be substantial disagreement between Kuhn and (for example) Price, it is generally agreed that highly significant discoveries which revolutionise conventional scientific wisdom and provide the opportunity for radical development of a particular field are the most uncertain and unreliable to predict (see, for example, Kuhn, 1970, pp. 92–110).

24. Jewkes et al. (1969, pp. 28–9, 153) mention the variety and difficulty of definition of development; the development of a revolutionary idea such as the jet or Wankel rotary engine, once the principle has been established, may require major commitments of resources and time, and by implication there may be significant uncertainty as to what will be required to put the radical new idea into commercial operation. On the other hand, development may mean no more than routine and systematic modification into a new form or style of a new product or process only marginally different from other existing products or processes.

25. It has been suggested that all major investments involve uncertainty to greater or lesser degrees and that R & D is only a form of investment albeit with a higher level of risk and uncertainty; see, for example, Machlup (1962a, pp. 188–92), Schumpeter (1954, pp. 88–90), Ansoff (1969, pp. 30–2) and Tilles (1966, pp. 190, 192). However, while replicability of investment in general poses no theoretical problem, replicability of discovery and invention *as* discovery and invention would be redundant unless the decision-maker was not in possession of the original specifications of the discovery or invention. Consequently there is an unavoidable qualitative difference between consideration of R & D inputs as investment and standard, potentially replicable, investment.

26. Links running from technology to science have been identified, particularly in the area of instrumentation (see, for example, Schmookler, 1962b, p. 197 *n*). However, while instrumentation may affect the precision with which information is found and the accessibility of new facts, it does not qualitatively affect the content of the discovery or information itself, but rather the manner in which the activity is conducted; it is factor rather than informational input.

27. See also Price (1965, p. 564).

28. For case histories of the development of nylon on which the discussion

above is based, see Mansfield (1968a, pp. 48–50) and Jewkes et al. (1969, pp. 275–7).

29. Jenner (1966, p. 792) suggests: '. . . it seems likely that most new operational features have occurred in clusters, rather than singly: the development of ball-point pens, for instance, introduced (at least) the operations of producing carbon copies of hand-writing and "instantly drying" ink.' A cluster effect in the applied research subsystem would emphasise that system's importance in transferring basic research to development, and further support the hierarchical interpretation. Scherer (1970, p. 356 n) in fact comments on Jewkes et al.'s series of case studies in this respect: 'The authors use a loose and sometimes curious definition of invention; many of the cases studied are best described as systems or conglomerations of inventions.' Examples which may be cited include the helicopter (pp. 257–60), television (pp. 307–10) and digital computers. While many subsequent products in each group might be classified under development, a number of innovations within each group required substantial applied research e.g. the autogyro, colour television and use of integrated circuits respectively.

Jenner's idea of clustering in terms of functional differentiation of new features with respect to a common source is therefore paralleled by frequently observed clustering of inventions within a given functional class. Jewkes et al. (1969, p. 171) do in fact identify the view that a scientific discovery may reveal a range of technical possibilities, while qualifying it with the comments that there may be substantial delay between discovery and applications, and that applications are often not obvious. The clustering of inventions suggests that, rather than basic research being independent of applied, applied research may be conducted without reference to *recent* basic research.

30. Defined as 'key contributions' (Sherwin and Isenson, 1967).

31. Project Hindsight appears to contradict the findings of TRACES and other studies mentioned earlier: undirected or non-mission-oriented research appears to play an insignificant role in development. However, Mansfield (1968a, p. 178) points out that this finding omits consideration of the 'pool of general knowledge' on which scientists and technologists may draw; also, since only post-war basic research events are included in the study, in the light of the typically long lag between basic research and subsequent applications mentioned earlier, the scope of the study may not be sufficient to put the contribution of basic research in perspective.

32. In 1970 industrial firms in the United States allocated $598 million to basic research, $3281 million to applied research and $13,978 million to development (National Science Foundation, 1972b, table 46, p. 83). These amount to 3, 18 and 78 per cent respectively of R & D funds (calculated to the nearest percentage).

33. Jewkes et al. (1969) analyse the growth of development costs using a definition of development similar to that of the N.S.F. (see p. 156). They suggest that development costs have escalated in the course of the twentieth century as accumulation of technical knowledge has widened development potential, and increasing sophistication on the part of the consumer has of necessity been paralleled by exhaustive testing and preparation, quality and reliability now being more immediate requirements compared to earlier periods of imperfect markets and undemanding consumers (pp. 155–60).

34. This is particularly liable to be the case with the N.S.F. scheme which

distinguishes the two research categories on the grounds of motivation. Gold (1971, p. 272 n) cites the 'circularity, inconsistency and variability' of reasons given for conducting research found in studies of the R & D function. Further, since the rational model of innovation is partly a result of *ex post* rationalisation (Schon, 1967, p. 37), we might reasonably expect a parallel *ex ante* rationalisation of intent in terms of high-sounding motives such as 'scientific knowledge for its own sake' as far as some marginal research activity is concerned. Machlup's analysis of qualitative differences between types of intended intermediate output is less open to ambiguous and discretionary interpretation on the part of respondents, though difficulties of definition and measurement still remain.

Chapter 3

1. See, for example, Blaug (1963), Fellner (1961) and Salter (1960).
2. Nelson et al. (1967, pp. 27–43, esp. p. 34 n.) conduct analysis of technological change activity in terms of supply and demand for invention.
3. For an example of how one eminent economist's work in this area has developed away from the neoclassical theory of the firm, see R. R. Nelson (1959a, 1972). In his earlier work Nelson described a rationally planned inventive effort as implying an excess of expected revenue over cost, while recognising the uncertainties typically found at the basic research end of the scale. Recently, however (1972), he has suggested that the formal, objectively rational, neoclassical theory of the firm is incapable of dealing adequately with the problems of R & D and innovation.
4. Measures of 'inventive activity' are more easily obtained than measures of propensity to innovate owing to the availability of patent statistics.
5. Article with Oswald Brownlee.
6. Schmookler does recognise the difficulties of referring specifically to determinate demand curves (1966, p. 184) but tends to understate the significance of uncertainty.
7. Standard Industrial Classification.
8. The work of Kenneth Arrow is emphasised in this section, partly because of his position as a leading theorist in decision-making under uncertainty, but also as a consequence of his attempts to relate such analysis directly to problems of technological change.
9. Emphasis added.
10. Mansfield (1971, p. 9) also interprets R & D as an activity aimed at reducing uncertainties.
11. By ignorance is meant unawareness, or simply lack of knowledge.
12. Menges (1968) comments: 'Uncertainty has *two* possible sources, not only random variation but also lack of knowledge concerning the possible states of nature or the distribution over the state space respectively' (p. 136). However, in the limiting case of perfect absence of knowledge of possible states of the world in a particular class (pure ignorance), uncertainty in the sense referred to above by Arrow ceases to be applicable.
13. Shackle (1970) puts his point strongly: 'What is novel is precisely that about which it is logically impossible for any statistical experience to exist' (p. 101). While it might be contended that novelty should e interpreted as a relative

rather than absolute term in the context of innovation, Shackle's criticism is suggestive of the difficulties involved in interpreting R & D as risky activity.

14. Arrow does recognise the 'fundamental paradox' in the determination of demand for information/invention: 'its value for the purchaser is not known until he has the information but then he has in effect acquired it without cost' (1962, p. 615).

15. Arrow does suggest that action may be decided on the basis of the average value of information/invention as a class in the past, which, if feasible, would by implication provide a basis for an objectively rational theory.

16. See Cyert and March (1963, pp. 36–8) for a full discussion of the nature of organisational slack.

17. For Cyert and March's full set of operational goals, see pp. 41–3.

18. Strategy here is interpreted in the same sense as used by Ansoff (1965a). He states: 'We use the term strategic to mean "pertaining to the relation between the firm and its environment"' (p. 18 n.). Ansoff points out that Cyert and March concern themselves only with a limited class of decision – operating decisions – and ignore strategic decision.

19. Ansoff defines operating decisions areas of concern as 'resource allocation (budgeting) among functional areas and product lines, scheduling of operations, supervision of performance, and applying control actions' (1965a, p. 18).

20. Search is 'simple-minded' in Cyert and March's view, similar or 'near' solutions being preferred (pp. 121–2). Minor differentation of products would tend to be cheaper, less uncertain, and take a shorter time to develop than more novel and radical innovations. However, this would not be interpreted as R & D in the N.S.F. categorisation; development work is concerned with converting research findings into actual products, not with differentiating an already existing product base (see National Science Foundation, 1973b, p. 19).

21. 'The "satisficing" model . . . seems inconsistent with the practice of highly profitable firms to continue to do considerable R & D; it does not adequately model in either the "switch" or the "search" sense the looking to other firms that seems to characterise "diffusion" processes, and it seems unable to account for "major" innovation' (Nelson, 1972, p. 42 n.).

22. See Ansoff (1965a, pp. 18–22).

23. Shubik (1970, p. 419) describes Cyert and March's work (often paraphrased as '*the* behavioural theory of the firm') as 'some behavioural theorising about the firm'. The indefinite article in the title suggests that they did not expect their work to represent the terminal point of theoretical development in this area, yet at the present time it is still accepted as the definitive text.

24. This interpretation of resource will be utilised in subsequent analysis.

25. Only limited evidence has been presented so far to suggest that R & D allocations are relatively stable. This will be discussed further in the next chapter.

26. Stimuli is interpreted as the set of actions which may be reported as autonomous from the point of view of analysis. Response is interpretable as pre-programmed reaction on the part of a system or systems.

27. The open system is often defined as engaging in exchanges with its environments, and compared in this respect to closed systems (see, for example, Emery and Trist, 1960). However, this tends to obscure the more subtle, but equally important, distinction between cybernetic and open systems.

28. We are restricting system definition to analysis and reorganisation of

information. It is of course possible to interpret all business enterprises as by definition being open systems in so far as they continuously import and export materials, resources and goods from their environments. However, restriction of system definition to informational levels will facilitate subsequent analysis.

29. This may account for the confusion over the status of cybernetic systems. For example, Kast and Rosenzweig (1974, p. 102) describe Boulding's third (homeostatic) level as a closed system, while Buckley (1968, p. 490) is typical of a number of other interpretations which describe the homeostatic system as an open system. Van Bertalanffy's analysis suggests that homeostatic mechanisms and subsystems are *characteristics* of open systems, and if this is accepted as describing their status in analysis of living systems, it may explain the source of the confusion; homeostatic systems alone do not fully define open systems, although they are typically integral features of mature open systems.

30. Problemistic search differs in other respects from normal homeostatic response. For example, the generation of corrective responses is typically a continuous process in homeostatic systems, while in the former case correction is by means of the discovery of discrete solutions; indivisibilities are therefore a feature of problemistic search output. Further, search may be terminated on perception of what is expected to be an adequate solution (Cyert and March, 1963, p. 120), and consequently anticipated future behaviour of the system may be sufficient to terminate search, while in normal homeostatic systems actual responses (e.g. furnace operations) are necessary to terminate the operations of the homeostatic mechanism.

31. See Cyert and March (1963, pp. 101–13) for a discussion and description of standard operating procedures.

32. For example, 'coping with potentially antagonistic objectives' might be dealt with by mechanisms involved in 'quasi-resolution of conflict' (see Cyert and March, 1963, pp. 116–18). In this case, sequential attention to objectives might be an operational means of handling potential conflict.

Chapter 4

1. See George (1974, pp. 33–6).

2. See Scherer (1970, pp. 41–4) and George (1974, pp. 36–7).

3. Chandler bases his analysis on historical data and detailed case histories of Du Pont, General Motors, Standard Oil and Sears Roebuck.

4. See Berle and Means (1932).

5. Scherer (1970, p. 30) states that the median firm in terms of shareholder numbers in the *Fortune* list of the 500 largest corporations for 1956 recorded more than 9000 shareholders.

6. Williamson (1970, 1971) uses this term to describe the functionally subdivided firm.

7. As a consequence of this growth strategy, Chandler identifies metals and a number of agricultural companies as being among those still operating the U-form in 1960 (1962, p. 51).

8. Thus, Galbraith (1967, pp. 72–82) talks of the desire and ability of large, diversified, management-controlled firms such as General Motors to control and stabilise their environment through planning. Schon (1971, p. 64) defines the

major planning question of the M-form enterprise as 'what are the potentials in development for new commercial ventures', and demonstrates how firms which integrate a number of diverse and related activities may unilaterally innovate whole new technological systems without waiting for complementary innovation on the part of other systems in the environment.

9. However, stability and predictability are achieved by rule-of-thumb decision rules (p. 120). While stability would appear to be an implicit element of a negotiated environment in conditions of uncertainty, it does not enter into Cyert and March's formal model.

10. Channon (1973, p. 4) cites evidence to suggest that approximately 86 per cent of the 500 largest manufacturing firms in the United States were administered by a multi-divisional structure in 1967.

11. See studies of two highly diversified M-form corporations, T.R.W. (Sheehan, 1966) and Texas Instruments (Layton, 1972, pp. 96–101), for examples of methodical formulation and integration of corporate strategy. Also, Carter (1971) in a field study of decisions relating to computer installation found that *active* search for opportunity paralleled problemistic search. Carter suggests that behavioural theory should be reappraised to include strategic considerations.

12. For example, Jantsch comments on the evolution of corporate planning in the specific context of rapid technological change (1969, p. 17).

13. Schon in fact emphasises this problem: 'The times required for diagnosis, for design of demonstration or for extension to the next instance, are long enough – in a period of loss of the stable state – to include changes which invalidate conclusions once they are reached' (1971, pp. 212–13).

14. Simon defines programmed decisions as being repetitive and routine to the extent that a definite procedure has been worked out for handling them. Unprogrammed decisions are such to the extent that they are novel and unstructured (1965, pp. 58–9). Strategy is interpretable as unprogrammed decision.

15. For examples and analysis of such control mechanisms, see Chamberlain (1962).

16. Ansoff (1967) states that operations research has been applied successfully to lower-level routine problems of internal allocation and utilisation of resources; 'variables which are exogenous to the company as a whole usually do not have an important effect on the solution' (p. 306). However: 'There is another large class of business problems in which operations research has had much less success. This includes company product policy, market policy, diversification, resource allocation, R & D planning, facilities location, etc. Problems in this class also have common characteristics. They are usually of concern to the top management levels in the firm. They deal with decisions whose impact is primarily on the long range future of the company and which affect the position of the company within its environment. Therefore variables and events exogenous to the firm frequently have a first order effect on the solution' (p. 306).

17. See Chandler (1962, pp. 382–3) and Williamson (1970, pp. 115, 117).

18. Consequently, relations with environment tended to parallel those of third-level mechanisms, with environmental stimuli of adversity being necessary to prompt corporate action, though analogous programmed responses were not always available. The firm became a reactive, adaptive system in the Cyert and March sense. The initial adoption of the M-form itself provides a good example

of environmentally induced corporate change, in effect distress innovation: 'In Du Pont, the company's financial statement . . . provided the shock that finally precipitated a major reorganisation. . . . At General Motors, an inventory crisis together with the collapse of the auto market in 1920 produced the change. . . . Partial reorganisation at Jersey was induced by excessive inventories, falling profits, and a declining market share. . . . But it was not until earnings fell . . . that major organisation changes were induced. . . . Although profit pressures at Sears were less dramatic, they also contributed to the change' (Williamson, 1970, p. 113; from Chandler, 1962, as source).

19. The open system property of differentiation is a separate concept from the economic concept of product differentiation, though, as is pointed out above, the latter encouraged the former in the development of the M-form.

20. See Von Bertalanffy (1973, p. 223) and Katz and Kahn (1966, p. 25).

21. See Jewkes et al. (1969, chap. 3) for examples and further discussion.

22. This, in fact, is a major contributory factor identified by Emery and Trist (1965).

23. Coupling is a term used to describe links facilitating the integration of subsystems in the overall system. It is particularly used to refer to the R & D subsystem and the development of technological innovation (see Jantsch, 1969, p. 24; Freeman, 1974, pp. 165, 169, 190–1; Price and Bass, 1969, p. 805).

24. For empirical analysis emphasising the importance of coupling, see Science Policy Research Unit (1972). For normative analysis of desirability of coupling, see Jantsch (1969).

25. See Williamson (1970, p. 114) for a fuller discussion of these problems in the large U-form organisation.

26. Thus Reeves (1958, p. 135) suggests that building up the research budget from a project base might be thought to be the ideal way of determining research expenditure. For examples of formal analysis and normative models of optimisation using project-based models, see Trattner and Zidaroiu (1972), Dean and Sengupta (1962), Gover and Srnavasan (1972) and Foster (1971).

27. However, Hamberg (1963, p. 110), while indicating that the R & D director usually has discretionary control over projects, points out the increasing influence and participation of senior management in this area. This is consistent with the post-war integrative phase of the R & D function identified by Fusfield (1975). At the same time, Nelson (1962) found that scientists at Bell Telephone Laboratories exercised substantial control over direction of R & D from within the function itself: 'If all allocation decisions were made by a centrally situated executive, the changing allocation of research effort called for as perceived alternatives and knowledge change would place an impossible information processing and decision making burden on top management.' Control and direction of the inherently uncertainty R & D projects is consequently made even more complex in terms of analysis of the decision-making process.

28. Nelson (1972) states that, as far as corporate behaviour is concerned, 'in an environment of rapid change where the lower-order rules may be quite unstable, one might hope to find more stability in the qualitative "meta-rules" that guide changes in the rules' (p. 42). Nelson gives as example of meta-rules, R & D style, search rules and broad strategies. In a similar vein, Freeman (1974, chap. 8) describes the various R & D strategies open to the firm, interpreting strategy as the broad R & D philosophy or approach of the firm, a functional equivalent of

the project-oriented concept of strategy discussed earlier. At higher levels in the firm it may be that such qualitative meta-rules may parallel the general stability in resource allocation described above.

29. Allen (1970, p. 176) found in his study that the research budget was typically set as a percentage of turnover or operating profit, while the prevalence of the convention of setting R & D as a percentage of sales is also identified by Mansfield (1968a, p. 62), Freeman (1974, pp. 245–6), Quinn (1959, p. 295) and Thomas (1963, p. 307).

30. In a survey of advertising appropriations, Taplin (1959, pp. 32, 235) found that the most common standard used was a 'fairly stable percentage of sales' (p. 232). Jastram (1949) reports a survey of 194 large advertisers in which three-quarters of firms reported restricting advertising outlays to a fixed percentage of sales. Doyle (1968) reports this as a consensus method of deciding appropriations, and sets it in the context of failure of rational, analytical method (pp. 576–7).

31. See also Von Bertalanffy (1973, pp. 130–5, 149–50, 164–9).

32. Argument partly based on N.S.F. survey (1956).

33. See also Dalton (1974, p. 145): 'technological innovation is no longer the haphazard result of occasional discovery. It has become institutionalised through corporate, university and governmental research.' Jewkes et al. (1969) also fell into the trap of the *non sequitur* in analysing progressive institutionalisation of R & D: 'The underlying principle rarely formulated precisely, but ever present, has been that originality can be organised . . . mass production will produce originality just as it can produce sausages' (p. 179). While such behaviour and attitudes may be concomitant with institutionalisation, they do not form necessary principles or rationale underlying institutionalisation *per se*.

34. This is paralleled in the problem of organisation structure by Burns and Stalker's (1961) distinction between mechanistic and organic systems of management. The organic system, typified by continual redefinition of tasks and shifting responsibility, is suggested to be the appropriate form for dynamic environment. Thus, while the organic form is identifiable as a coherent *system* composed of persisting organisation goals, resources, etc., its content and distribution of activity is liable to be dynamic, non-routine and highly changeable.

35. General business uncertainty is also aggravated by the development of turbulent fields, in this case in the macro-environment.

36. This refers to both organisational and technological connections and links between component subsystems in the firm. See Gold (1971, p. 222) for evidence of direct relationships between radical innovation and tendency to substantially readjust existing organisational arrangements and functional allocations, and Glennan (1967, pp. 32–4) for examples of problems presented by component interrelatedness, for technological innovation.

37. See Jewkes et al. (1969, pp. 115–16) for examples of myopia and underestimation of outside ideas on the part of some eminent researchers.

38. In such circumstances it is perhaps surprising that search for major innovation is conducted at all. However, despite the fact that only about 3 per cent of R & D expenditure was allocated to the basic research subsystem in the N.S.F. categorisation (N.S.F., 1973b, from table B-45, p. 66), the distribution of

scientists to basic research work activities was approximately 17 per cent of all non-managerial employment of scientists in R & D subsystems (N.S.F., 1973a, from chart, p. 13). While basic research tends to cost relatively little in terms of demands on resources, development costs typically involve expensive prototypes and pilot projects. Such expenditure and costs are incurred in development and tend to overstate the activity in that subsystem and understate activity in the basic research subsystem, in terms of the allocation of scientific time and work.

Chapter 5

1. This point may appear trivial, yet there is evidence to suggest that its significance is not generally comprehended. Mueller (1967) states with respect to his analysis: 'emphasis is upon the complexity of this [corporate] behaviour, and upon the eventual need for attempting to explain this behaviour with models of corresponding complexity' (p. 58). His analysis constitutes an econometric study of the firm's decision process, and if Simon's hypothesis generalises to organisations, at least some complex behaviour may be simulated using simple models of decision and variable input (the latter corresponding to turbulent fields).

2. See, for example, 'Flow diagram of world' (in Forrester, 1972, p. 207). A good example of an accurate, highly abstract model of a system which is extremely complex when analysed microscopically, is the second-level clockwork model of the solar system.

3. 'The important distinction between resources and services is not their relative durability; rather it lies in the fact that resources consist of a bundle of potential services and can, for the most part, be defined independently of their use, while services cannot be so defined' (Penrose, 1959, p. 25).

4. See this article for full argument and analysis.

5. In an explicit systems context, this is paralleled by the co-ordination, control and direction of maintenance subsystems and adaptive subsystems in the organisation (Katz and Kahn, 1966, p. 39).

6. For examples and analysis of extrapolative technological forecasting based on such perceiving or hypothesised patterns, see Martino (1972), Cetron (1969) and Ayres (1969).

7. See also Katz and Kahn (1966, p. 22) and Von Bertalanffy (1973, pp. 42–4).

8. This latter problem is the area studied by the behavioural theory of the firm. Action contingent on deviation from steady state is analysed, but not the determinants of the steady state itself (see Cyert and March, 1963, pp. 99–101).

9. 'A collection of physical things can always be treated as if they were divisible into units of a single commodity so long as their relative prices can be assumed unchanged, in the particular problem in hand. So long as the prices of other consumption goods are assumed to be given, they can be lumped together into one commodity "money" or "purchasing power in general". Similarly in other applications, if change in relative wages is to be neglected, it is quite legitimate to assume all labour homogenous' (Hicks, 1946, p. 33).

10. We would expect substitutability of resources between functions to be low in the short run; R & D scientists do not become salesmen because of short-run changes in preferences. The time period referred to above would generally

operate over a number of calendar years, providing both time to obtain utility from resource allocations in long horizon functions, such as R & D and marketing, and opportunities to run down, build up or otherwise adjust function allocations as required.

11. This argument based on 'natural selection' of survivors is similar to that developed by Alchian (1950) to justify treating typical (surviving) firms as relatively more efficient than non-survivors. However, the external pressure of turbulent fields, directional information as to desirable system design supplied by feedback, and resource orientation of the above model, all constitute significant differenccs from Alchian's argument. Alchian's argument has been criticised from the point of view of lacking a mechanism corresponding to genetic inheritance in the Darwinist model of evolution (see Penrose, 1952; Winter, 1964, 1971). In the model above, informational feedback is a mechanism whose function corresponds to genetic inheritance; however, the mechanism operates through managerial learning and corresponding resource adjustment, unlike the non-purposive genetic inheritance mechanism. Through interaction with its environment, the firm builds up a Gestalt of the implications of different combinations of subsystems for corporate performance and viability and *adapts* allocations accordingly. Genetic inheritance of surviving characteristics constitutes a less flexible adaptation mechanism owing to the fixed structural arrangements of organisms.

12. The framework permits analysis of the effects of changes in relative composite resource price *between* functions assuming constant relative prices *within* functions, and stable preferences. Changing relative prices and incomes in a stable preference system is a central concern of consumer theory, but in fact our concern would be in effect the reverse of this; our preoccupation will be mainly with changes in preferences in the context of constant relative prices and budget constraints.

13. See also Katz and Kahn (1966, pp. 25–6).

Chapter 6

1. The wording of this technique in Näslund and Sellstedt's questionnaire was: 'Analysis of individual projects. Each project is analysed according to its own potential of becoming successful. A predetermined budget does not exist' (p. 72).

2. The total number of times all methods were used add up to more than 69 owing to firms employing more than one method in deciding their budget. Larger firms tended to use more than one method to a greater extent than smaller firms (Näslund and Sellstedt, 1974, p. 68).

3. Bloom qualifies his description of the project-based budgeting procedure by suggesting that management must have some guide to how many projects will be approved: 'Probably a rough figure is arrived at by looking at other companies' expenditures and sometimes by using a percentage of sales. The total funds requested by the various departments may add up to more or less than this figure, which then must be revised up or down' (p. 610). However, it is not clear how this 'guide' affects the budget decision, since it is revised according to aggregated project appropriations, not vice versa.

4. However, Smith and Creamer point out that even in small companies existing cost of establishing R & D tended to be used as a base line for calculating next year's budget (pp. 135, 142). It would be unfair to suggest that Näslund and Sellstedt unqualifiedly recommend adoption of optimisation techniques; in their 1974 article, advantages and disadvantages are given for each method primarily in terms of information costs and, in the case of rational project models, the conservative bias of such techniques.

5. A further circumstance which would probably mitigate against the existence of a stable resource preference system in the immediate post-war period is the war itself. The reorientations of production and research from a war-time to a peace-time footing would radically alter conditions under which R & D operated. Even if a stable resource preference system operated before or during the war, we would expect a period of instability and re-evaluation in the post-war period during which management adjusted to the changing circumstances.

6. '"Big Science" involves a research system in which a consciously articulated goal exists . . . [and] also implies a system in which there has been a commitment of resources and the organisation or co-ordination of skills and institutions on a scale which only national governments can undertake' (Blankenship, 1974, p. 257).

7. For projects of a given size, condition (ii) might also be vulnerable in group (a) owing to the smaller R & D programmes conducted. However, if, as has been suggested, the scale of individual projects was typically much less than present-day projects (such as prevalence of big science projects in the latter case), this would be a compensatory factor. Nevertheless, in general, the further back in time that institutionalised R & D decision-making is studied, the more vulnerable we would expect condition (ii) to become.

8. Analogously, individuals may recognise and discriminate between faces, yet not be able to 'explain' recognition. In certain uses, conspicuous features or components of the 'Gestalt' may facilitate recognition, yet conscious articulation of reasons for discrimination may not be generally possible, especially if rational explanation is expected on the basis of comparative analysis of facial features.

Chapter 7

1. See 'Technical Notes' in N.S.F. (1966, pp. 159–60).

2. However, see Wu and Pontney (1967, pp. 54–5) for an analysis of the effect of a taste parameter on consumer preferences.

3. Expanding the terms within brackets, $\mathrm{d}S_1/\mathrm{d}\alpha$ is the sign of:

$$\frac{\partial}{\partial\alpha}\left(\frac{\partial U}{\partial S_1}\right)-\frac{\partial}{\partial\alpha}\left(\frac{\partial U}{\partial S_2}\right)$$

4. In this context, 'functional' pertains to behaviour and characteristics of derived final output, not the overall system (such as the R & D 'function').

5. In Phillips' definition, P_j is a discrete variable taking the values 1, 2 and 3. Here we assume that it is continuous.

6. A striking feature of this interpretation of synergy is its similarity to the simplistic (and misleading) definition of 'Gestalt' as 'the whole is greater than the

sum of the parts'. Indeed, much of Ansoff's analysis takes place at a high level of abstraction (see, for example, table 5.1, p. 77), though he uses the concept both to describe relationships in 'functional areas' (such as R & D, marketing) at resource level *and* at project level.

7. Behavioural theory might suggest other relations between growth and R & D. Low growth might signal distress conditions and stimulate search for new products – R & D as problemistic search. Alternatively, successful growth may generate slack resources over and above that required for survival of the corporation, and encourage long-run R & D projects and resource allocation that would not be sanctioned if the survival of the firm was threatened. However, from our earlier argument, this is essentially *non*-steady-state behaviour with which our model is not designed to deal. We assume survival to be indefinitely marginally feasible for the corporation, and that neither slack nor distress conditions are applicable.

8. 'Qualified Scientists and Engineers'.

9. See, in particular, Comanor, (1967), Scherer (1965) and Mansfield (1964). Jewkes et al. (1969) conclude that there is no simple relationship between size of firm and research intensity.

10. Whereas R & D as dependent variable was interpreted in terms of company-financed activity, the above argument relates to total financing of basic research. This is because the N.S.F. does not provide separate figures for company-financed basic research, but includes both federal and company funds together when analysed by industry and size of company (see N.S.F., 1966, table A-69, p. 140).

11. See the argument earlier in the chapter for justification of cross-section analysis.

12. The U.S. Census of Manufactures (see Bureau of the Census, 1968b) is conducted on a five-yearly basis, indices of production being provided on this basis, the preceding census providing the base for calculating change in production. A further variable utilised in the analysis of this chapter is V_j, obtained from vol. III of the Census for 1963, and again this is a further reason for choosing 1963 as the year for cross-sectional analysis.

13. Here lower- and higher-order refers to the place of a particular hypothesis in the overall theoretical structure, in the sense meant by Cyert and Grunberg (1963), in Cyert and March (1963), and referred to earlier.

14. Net sales is recorded dollar value for goods sold or services rendered less returns, allowances, freight charges and excise taxes (see Appendix III for full details).

15. Value added equals value of shipments less cost of materials, supplies, containers, fuels, purchased electricity and contract work plus net change in finished goods and work-in-progress inventory and value added in merchandising activities of manufacturing establishments.

16. The figures were abstracted from the earlier comprehensive report on the 1963 census also compiled by the Bureau of the Census (1968a).

17. V_j was also not available for the ordnances sub-industry, which is classified along with aircraft and missiles (see Appendix III, 'Explanation of tabular data'). The V_j for aircraft was therefore taken as an approximation since it accounted for 73 per cent of the value added of this sector in 1963 (current dollars) (from Bureau of the Census, 1968b).

18. Since R & D as a fraction of sales only ranges from almost zero in some cases to 0.055 of net sales in the largest size category of the instruments industry, while V_j ranges from 0.21 (petroleum) to 0.76 (drugs), we would not expect to find R & D expenditure having a significant effect on the pattern and values of V_j in different industries (see Appendix I, columns (1) and (2). If we assume that value added/value of shipments does not vary significantly for either value of *i* in each industry, then V_j may be used to adjust R_{ij}/S_{ij} to provide estimates of X_{ij}. This would be the case if V_j was determined by technological factors independent of the size of firm for a particular industry.

19. For further analysis of the respective functional forms, see Johnston (1963, pp. 47–50).

20. Dr. Swenson is a U.S. citizen who previously worked with NASA as an engineer before joining the Industrial Science Department at Stirling. He was requested to estimate P_j for the remaining industries for the time period of the analysis, using Phillips' definition and estimates for the other industries as a guide.

21. The importance of wages and salaries of scientists and engineers as a proportion of R & D cost was initially measured in two ways in the original data matrix. By definition, the two measure U_j and U'_j sum to 100 per cent, the difference being that we would expect a negative relationship between U'_j and X_{ij}, U'_j and B_{ij}.

22. The convention adopted in this sample and others involving Dr. Swenson's estimates of P_j is that sub-industries are used if values for each variable are available for at least two sub-industries. Otherwise, values for the overall industry are used (if available), and then a sub-industry set of estimates (if available).

23. As indicated from the data matrix in Appendix I, further degrees of freedom were obtainable owing to the existence of more estimates for T_{ij}/S_{ij} than for R_{ij}/S_{ij}. However, to facilitate comparison with the X_{ij} estimating equations, sample (d) is still used.

24. It is difficult to assess to what extent this 'conventional wisdom' is a consequence of observed behaviour, and to what extent it is a result of the distrust held by many U.S. authorities for federal involvement in private enterprise. However, for some indication of the latter in this context, see the quote from Boundy and Chamberlain (1959, p. 83) cited earlier.

25. Further support for Collier's observation is the highly publicised widespread unemployment of Q.S.E.s following the cutback in U.S. defence expenditure in the 1960s, and the phasing-out of the large-scale U.S. space programmes following the manned moon landing.

26. One possibility is that Q.S.E. intensity does not directly affect R & D intensity (and consequently is inappropriately specified as an environmental variable), but is itself a dependent variable in the resource allocation process; since the variable measured at corporation level would refer to the distribution of resources within the R & D function, it may be that future analysis should first of all be concerned with determinants of Q.S.E. intensity rather than any possible role it may have in directly influencing resource allocation.

27. A further implication of two of the functional forms, the linear and recip. transform., is the possible existence of threshold values. However, we cannot identify a single threshold value in a multiple regression analysis, since a

threshold value for, say, size of firm would depend on the values of the other variables in the regression. Specific threshold values for R & D with respect to a particular independent variable must be determined by holding the values of the other variables constant.

28. A possible reason for the negative sign of U_j is that the science-based industries are more capital-intensive in their R & D than the less progressive industries, *ceteris paribus*. In this case, capital intensity of R & D might reflect the extent of the evolution of a scientific base in a particular industry, and be an alternative measure of scientific progressiveness to P_j.

29. As well as additional rows being provided for the regression equations by the extra estimates of P_j provided by Dr. Swenson, the two estimated for lumber were deleted since fitting doub. log. and log. recip. transformations would have involved taking logarithms of zero. This undoubtedly adversely affected the fit of the regression analysis, since the P_j, $G_{j'}$ and E_{ij} and E'_{ij} variables were all relatively low for lumber in both size categories, consistent with the expectations of the hypotheses.

30. A more general treatment would recognise that R & D itself might also be generally a lower-ranking preference to P/M activity and resources. We assume, however, that the firm has progressed beyond this stage and that the institutionalisation of R & D as a function is evidence of its full integration in the managerial preference system for resources.

31. The equations using G_j and E'_{ij} ((8.2), (8.4), (8.6) and (8.8)) had corrected R^2 values ranging from 0.46 to 0.79. G_j was significant in (8.2) and (8.4), E'_{ij} significant at the 99 per cent level of confidence in each equation.

32. Phillips (1966) conducts an analysis which is more nearly comparable to the analysis of this chapter, since R & D expenditure as a percentage of value added is regressed on size of firm in terms of value added, as well as other variables using N.S.F. data for 11 industries in 1958. No significant relationship was found, but this may be a consequence of multicollinearity; for example, size of firm and concentration ratio had a correlation coefficient of 0.81.

33. Comanor found, in his 1967 analysis, that moderate entry barriers are associated with level of research effort, after allowing for other variables. The relationship of this alternative measure of market power (or protection) to research effort was interpreted as suggesting that some protection may facilitate R & D, but that too much may lead to diminishing technological progressiveness.

34. This does not of course mean that discretionary behaviour may be necessarily unimportant in corporate decision-making. Multicollinearity with other independent variables may obscure the effect of concentration on research activity, or it may be that concentration is an inferior proxy for monopoly power.

Chapter 8

1. Innovation or inventive activity has been variously interpreted as patenting activity, R & D employment and total expenditure on R & D resources, among other definitions. As Mueller (1966) points out, these may be closely related and substitutable indicators of innovative activity. Innovative activity is therefore interpreted here as a broad concept for which different analysts have employed different operational definitions.

2. Schumpeter (1943) suggested that innovative activity was suited to conditions of monopoly power, and stimulated subsequent discussion and analyses on the relationship between size and/or concentration on innovative activity.

3. For examples of oligopolistic rivalry models of firm behaviour, see Kamien and Schwartz (1972a, b, 1974) and Grabowski (1970); Needham (1975) also suggests that monopolistic conditions may encourage imitative behaviour (p. 248). For oligopolistic interpretations in which rival firms match project development time or other variables, see in particular Baldwin and Childs (1969), Grabowski (1970), Horowitz (1963) and Scherer (1967).

4. The *Fortune* 500 is an annual listing of the 500 largest corporations in the United States for the particular year.

5. The Howe and McFetridge study was not available until the draft of this chapter was nearing final form. It is therefore useful in that it provides crude support for the adaptive learning interpretation with regard to expectations as to how R^2values would differ for the industries sampled.

6. A further source of error in both cases may be measurement error in the variables. It is implicitly assumed here and in Grabowski's analysis that is not important.

7. A further study by Globerman (1973) investigated the determinants of research intensity (measured as R & D personnel as a proportion of all employees) in 15 Canadian manufacturing industries for 1965–9. Two samples were used for his regression analysis, one utilising 9 'technologically progressive' industries according to the Scherer index, the other using 6 'unprogressive' industries. Three independent variables were used in both regressions: a measure of foreign-held industry assets, a concentration index, and a measure of government subsidy to R & D. In the technologically unprogressive case all variables were insignificant even at the 0.10 level, while in the technologically progressive case all variables were significant at the 0.01 level. The R^2 values were 0.175 and 0.810 respectively. While this may be cited as further support of the argument presented in this section, it is not presented in the main text because of the statistically weak nature of Globerman's analysis, particularly the few degrees of freedom obtained in each regression.

Chapter 9

1. It might be objected that the smallest constituents are the household and the firm, respectively. However, since the concepts utilised in consumer theory are those describing choice behaviour of the individual consumer, and modern development of the theory of the firm has been extended to cover multi-product firms whose decision-making is based on optimising project allocation, this criticism may be disregarded.

2. Koopmans' emphasis.

References

Adams, W. J. (1970). Firm size and research activity: France and the United States. *Quarterly Journal of Economics*, 84, 386–409.

Alchian, A. A. (1950). Uncertainty, evolution and economic theory. *Journal of Political Economy*, 58, 211–21.

Allen, J. M. (1970). A Survey into the R & D evaluation and control procedures currently used in industry. *Journal of Industrial Economics*, 18, 161–81.

Allport, F. H. (1955). *Theories of Perception and the Concept of Structure*, Wiley, New York.

Ames, E. (1961). Research, invention, development and innovation. *American Economic Review*, 51, 370–81.

Angyal, A. (1969). A logic of systems. In: F. E. Emery (ed.), *Systems Thinking*, Penguin Books, London, pp. 17–29.

Ansoff, H. I. (1965a). *Corporate Planning*, Penguin Books, London.

Ansoff, H. I. (1965b). The firm of the future. *Harvard Business Review*, Sept–Oct, pp. 162–78.

Ansoff, H. I. (1967). A quasi-analytic method for long-range planning. In: B. V. Carsberg and H. C. Edey (eds.), *Modern Financial Management*, Penguin Books, London, pp. 303–22.

Ansoff, H. I. (1969). Towards a strategic theory of the firm. In: H. I. Ansoff (ed.), *Business Strategy*, Penguin Books, London, pp. 11–40.

Aram, J. D. (1973). Innovation via the R & D underground. *Research Management*, 16, 24–6.

Arrow, K. J. (1952). Economic welfare and the allocation of resources for invention. In: National Bureau of Economic Research (1962) pp. 609–26

Arrow, K. J. (1968). Mathematical models in the social sciences. In: M. Brodbeck (ed.), *Readings in the Philosophy of the Social Sciences*, Macmillan, New York, pp. 635–67.

Arrow, K. (1971a). Classificatory notes on the production and transmission of technical knowledge. In: K. Arrow, *Essays in the Theory of Risk-Bearing*, Markham Publishing Co., Chicago, pp. 165–76.

Arrow, K. (1971b). Exposition of theory of choice under uncertainty. In:

K. Arrow, *Essays in the Theory of Risk-Bearing*, Markham Publishing Co., Chicago, pp. 44–89.

Ashby, W. R. (1960). *Design for a Brain*, 2nd ed., Wiley New York.

Ayala, F. J. and Dobzhansky, T. (1974). *Studies in the Philosophy of Biology: Reduction and Related Problems*, Macmillan, London.

Ayres, R. U. (1969). *Technological Forecasting and Long-Range Planning*, McGraw-Hill, New York.

Baldwin, W. L. and Childs, G. L. (1969). The fast second and rivalry in research and development. *Southern Economic Journal*, 36, 18–24.

Barbour, I. G. (1966). *Issues in Science and Religion*, S.C.M. Press, London.

Barnard, C. I. (1948). *The Functions of the Executive*, Harvard University Press, Cambridge, Mass.

Baumol, W. J. (1959). *Business Behaviour, Value and Growth*, Macmillan, New York.

Becker, M. (1974). Reduction, hierarchies and organicism. In: Ayala and Dobzhansky (1974) pp. 163–76.

Ben-David, J. (1972). Scientific entrepreneurship and the utilization of research. In: B. Barnes (ed.), *Sociology of Science*, Penguin Books, London, pp. 181–7.

Berle, A. A. and Means, G. C. (1932). *The Modern Corporation and Private Property*, Macmillan, New York.

Bernal, J. D. (1939). *The Social Function of Science*, Macmillan, London.

von Bertalanffy, L. (1968). General System Theory–a critical review, in W. Buckley, *Modern Systems Research for the Behavioural Scientist*, Aldine, Chicago.

von Bertalanffy, L. (1973). *General System Theory*, Penguin University Books, London.

Black, G. (1969). *The effect of government funding.* In: Gruber and Marquis (1969) pp. 179–218.

Blackett, P. M. S. (1968). Memorandum to the Select Committee on Science and Technology. *Nature* (*London*), 219, 1107–10.

Blankenship, L. V. (1974). Management, politics and science: a nonseparable system. *Research Policy*, 3, 244–57.

Blaug, M. (1963). A survey of the theory of process innovations. *Economica*, 30, 13–32.

Bloom, G. F. (1951). Union wage pressure and technological discovery. *American Economic Review*, 12, 5–24.

Bogaty H. (1969), Employees of technical organisations can be motivated. *Research Management*, 12, 524.

Borch, K. (1968). Introduction. In: K. Borch and J. Mossin (eds.), *Risk and Uncertainty*, Macmillan, London, pp. xiii–xv.

Boulding, K. (1956). *The Image*, University of Michigan Press, Ann Arbor.

Boulding, K. (1968). General systems theory – the skeleton of a science in W. Buckley, *Modern Systems Research for the Behavioural Scientist*, Aldine, Chicago, pp. 3–12.

Boundy, R. H., and Chamberlain, Jr., C. H. (1959). Problems and prospects of industrial research. *Research Management*, 2, 81–96.

Bower, J. L. (1970). Planning within the firm. *American Economic Association, Papers and Proceedings*, 60, 186–94.

Bowie, R. M. (1963). The direction and control of research and development. *Research Management*, 6, 277–88.

Bradbury, F. R. (1974). Aspects of process development in the fine chemicals industry: some principles of process development. *Chemistry and Industry*, May, pp. 392–6.

Branch, B. (1974). Research and development activity and profitability: a distributed lag analysis. *Journal of Political Economy*, 82, 999–1011.

Braybrooke, D. and Lindblom, C. E. (1963). *A Strategy of Decision*, Free Press, New York.

Brown F. C. (1962). The impact on business firms II. In: Symposium on the Impact of Federal Government Research and Development on Industrial Research and on the Economy of the United States. *Research Management*, 5, 341–59.

Brozen, Y. (1962). The future of industrial research and development. In: National Bureau of Economic Research (1962) pp. 273–6.

Brozen Y. (1965). R & D differences among industries. In: R. A. Tybout (ed.), *Economics of Research and Development*, Ohio State University Press, Columbus, pp. 83–100.

Bryan, G. L. (1973). The role of basic research in the total R & D process. *Research Management*, 16, 29–33.

Buckley, W. (1968). Society as a complex adaptive system. In: W. Buckley (ed.), *Modern Systems Research for the Behavioural Scientist*, Aldine Publishing Co., Chicago, pp. 490–513.

Bureau of the Census (1968a). *Census of Manufactures 1963*, vols. I and III, U.S. Dept. of Commerce, Bureau of the Census, Washington.

Bureau of the Census (1968b). *Census of Manufactures 1963*, vol. IV: *Indexes of Production*. Federal Reserve Board, Washington.

Bureau of the Census (1970). *Statistical Abstract of the United States 1970*, 91st ed., U.S. Dept. of Commerce, Bureau of the Census, Washington.

Burns, T. and Stalker, G. M. (1961). *The Management of Innovation*,

Tavistock Publications, London.

Carter, C. F. and Williams, B. R. (1957). *Industry and Technical Progress*, Oxford University Press, London.

Carter, C. F. and Williams, B. R. (1958). *Investment and Innovation*, Oxford University Press, London.

Carter, C. F. and Williams, B. R. (1959). *Science in Industry: Policy for Progress*, Oxford University Press, London.

Carter, E. E. (1971). The behavioural theory of the firm and top level decisions. *Administrative Science Quarterly*, 16, 413–28.

Cetron, J. M. (1969). *Technological Forecasting*, Gordon & Breach, New York.

Chamberlain, N. W. (1962). *The Firm: Micro-economic Planning and Action*, McGraw-Hill, New York.

Chamberlain, N. W. (1968). *Enterprise and Environment: The Firm in Time and Place*, McGraw-Hill, New York.

Chandler, Jr., A. D. (1962). *Strategy and Structure*, Doubleday, New York.

Channon, D. F. (1973). *The Strategy and Structure of British Enterprise*, Macmillan, London.

Clare, J. W. H. (1963). Current trends in the organisation of industrial research. *Research Management*, 6, 135–52.

Collier, D. W. (1963). A 'civilian' looks at government-sponsored R & D. *Research Management*, 6, 335–46.

Collier, D. W. (1964). How should management determine how much company funds to invest in R & D? *Research Management*, 7, 393–406.

Comanor, W. S. (1965). Research and technical change in the pharmaceutical industry. *Review of Economics and Statistics*, 47, 182–90.

Comanor, W. S. (1967). Market structure, product differentiation and industrial research. *Quarterly Journal of Economics*, 81, 645–52.

Cyert, R. M. and Grunberg, E. (1963). Assumption, prediction and explanation in economics. In: Cyert and March (1963) pp. 298–311.

Cyert R. M. and March, J. G. (1963). *A Behavioural Theory of the Firm*, Prentice-Hall, Englewood Cliffs, N. J.

Dalton, G. (1974). *Economic Systems and Society: Capitalism, Communism and the Third World*, Penguin Books, London.

Day, R. H. (1967). Profits, learning and the convergence of satisficing to marginalism. *Quarterly Journal of Economics*, 81, 302–11.

Dean, B. V. (1968). *Evaluating, Selecting and Controlling R & D Projects*, American Management Association Inc., A.M.A. Research Study, No. 89.

Dean, B. V. and Sengupta, S. S. (1962). Research budgeting and project selection. *IRE Transactions in Engineering Management*, EM 9, 158–69.

Dörfer, I. N. H. (1974). Science and technology in Sweden: the Fabians versus Europe. *Research Policy*, 3, 134–55.

Douds, C. F. (1971). The state of the art in the study of technology transfer – a brief survey. *R & D Management*, 1, 125–31.

Doyle, P. (1968). Economic aspects of advertising: a survey. *Economic Journal*, 78, 570–602.

Elliott, J. W. (1971). Funds flow vs. expectational theories of research and development expenditures in the firm. *Southern Economic Journal*, 37, 409–22.

Ellis, I. T. (1962). The impact on the economy as a whole. In: Symposium on the Impact of Federal Government Research and Development on Industrial Research and on the Economy of the United States. *Research Management*, 5, 363–75.

Emery, F. E. (1969). F. E. Emery (ed.), *Systems Thinking*, Penguin Books, London.

Emery, F. E. and Trist, E. L. (1965). The casual texture of organisational environments. Reprinted in Emery (1969), pp. 241–57.

Emery, F. E. and Trist, E. L. (1960). Socio-technical systems. Reprinted in Emery (1969), pp. 281–96.

Encel, S. (1970). Science, discovery and innovation: an Australian case history. *International Social Science Journal*, 42–53.

Falk, C. E. (1973). An operational, policy-oriented research categorisation scheme. *Research Policy*, 2, 186–202.

Fellner, W. (1961). Two propositions in the theory of induced innovations, *Economic Journal*, 71, 305–8.

Forrester, J. W. (1972). Understanding the counterintuitive behaviour of social systems. In: J. Beishon and G. Peters (eds.), *Systems Behaviour*, Harper & Row, London, pp. 206–17.

Foster, R. N. (1971). Estimating research payoff by internal rate of return method. *Research Management*, 14, 27–43.

Freeman, C. (1962). Research and development: a comparison between British and American industry. *National Institute Economic Review*, May, pp. 21–39.

Freeman, C. (1974). *The Economics of Industrial Innovation*, Penguin Books, London.

Friedman, M. (1953). *Essays in Positive Economics*, University of Chicago Press, Chicago.

Fuchs, V. R. (1972). *Policy Issues and Research Opportunities in Industrial Organisation*, National Bureau of Economic Research 50th

Anniversary Colloquium III, Columbia University Press, New York.

Fusfield, D. (1975). Industrial research – four stages of progress. *Research Management*, 18, 13–14.

Galbraith, J. K. (1967). *The New Industrial State*, Hamish Hamilton, London.

Gambo, G. C. (1959). Representative costs of doing research and development. *Research Management*, 2, 33–43.

George, K. D. (1974). *Industrial Organisation: Competition, Growth and Structural Change in Britain*, 2nd ed., Allen & Unwin, London.

Gibbons, M. and Johnston, R. (1974). The roles of science and technological innovation. *Research Policy*, 3, 220–42.

Gibson, R. E. (1962). A systems approach to research management I. *Research Management*, 5, 215.

Glatt, E. and Shelly, M. W. (eds.) (1967). *The Research Society* Gordon & Breach, New York.

Glennan, Jr., T. K. (1967). Issues in the choice of development policies. In: Marshak et al. (1967) pp. 13–48.

Globe, S., Levy, G. W. and Schwartz, C.M. (1973). Key factors and events in the innovation process. *Research Management*, 16, 8–15.

Globerman, S. (1973). Market structure and R & D in Canadian manufacturing industries. *Quarterly Review of Economics and Business*, 13, 59–67.

Gold, B. (1971). *Explorations in Managerial Economics: Productivity, Costs, Technology and Growth*, Macmillan, London.

Gordon, R. A. (1961). *Business Leadership in the Large Corporation*, University of California Press, Berkeley.

Gover, D. P. and Srnavasan, V. (1972). Allocating resources between research and development: a macroanalysis. *Management Science*, 18, 492–501.

Grabowski, H. G. (1968). The determinants of industrial research and development: a study of the chemical, drug and petroleum industries. *Journal of Political Economy*, 76, 292–306.

Grabowski, H. G. (1970). Demand shifting, optimal firm growth and rule of thumb decision-making. *Quarterly Journal of Economics*, 84, 217—36.

Grabowski, H. G. and Baxter, N. D. (1973). Rivalry in industrial research and development: an empirical study. *Journal of Industrial Economics*, 21, 209–35.

Grabowski, H. and Mueller, D. (1970). Industrial organisation: the role and contribution of econometrics. *American Economic Review, Papers and Proceedings*, 60, 100–4.

Green, H. A. J. (1971). *Consumer Theory*, Penguin Books, London.

Griliches, Z. (1958). Research costs and social returns: hybrid corn and related innovations. *Journal of Political Economy*, 66, 419–31.

Griliches, Z. (1962). 'Comment' on 'The origins of the basic inventions underlying Du Pont's major product and process innovations 1920 to 1950'. In: National Bureau of Economic Research (1962) pp. 346–58.

Gruber, W. H. and Marquis, D. G. (eds.) (1969). *Factors in the Transfer of Technology*, M.I.T. Press, Cambridge, Mass.

Hamberg, D. (1963). Invention in the industrial research laboratory. *Journal of Political Economy*, 71, 95–115.

Hamberg, D. (1966). *R & D Essays on the Economics of Research and Development*, Random House, New York.

Havrilesky, T. (1972). The information explosion, technological innovations and the competitive ethic. *Land Economics*, 48, 347–56.

Hicks, J. R. (1946). 2nd edn, *Value and Capital*, Oxford University Press, London.

Hirschman, A. O. (1958). *The Strategy of Economic Development*, Yale University Press, New Haven.

Hirschman, A. O. (1967). *Development Projects Observed*, Brookings Institution, Washington.

Hirschman, A. O. and Lindblom, C. E. (1962). Economic development, research and development, policy making: some converging views. *Behavioural Science*, 7, 211–22.

Hogan, T. and Chirichiello, J. (1974). The role of R & D in small firms. *Research Management*, 17, 26–30.

Holloman, J. H. (1965). Science and innovation. In: R. A. Tybout (ed.), *Economics of Research and Development*, Ohio State University Press, Columbus, pp. 251–7.

Horne, C. F. (1962). The impact on business firms I. In: Symposium on the Impact of Federal Government Research and Development on Industrial Research and on the Economy of the United States. *Research Management*, 5, 325–38.

Horowitz, I. (1962). Firm size and research activity. *Southern Economic Journal*, 28, 298–301.

Horowitz, I. (1963). Research inclinations of a Cournot oligopolist. *Review of Economic Studies*, 30, 128–30.

Howe, J. D. and McFetridge, D. G. (1976). The determinants of R & D expenditures. *Canadian Journal of Economics*, 9, 57–71.

Illinois Institute of Technology (1968). *Technology in Retrospect and Critical Events in Science*, vol. I, National Science Foundation, Washington.

Jantsch, E. (1969). Technological forecasting in corporate planning. In: G. Wills, D. Ashton and B. Taylor (eds.), *Technological Forecasting and Corporate Strategy*, American Elsevier Publishing Co., New York, pp. 17–38.

Jastram, R. W. (1949). Advertising outlays under oligopoly. *Review of Economics and Statistics*, 31, 106–9.

Jastram, R. W. (1950). The development of advertising appropriation policy, *Journal of Business*, 23, 154–66.

Jenner, R. A. (1966). An information version of pure competition. *Economic Journal*, 76, 786–805.

Jewkes, J., Sawers, D. and Stillerman, R. (1969). *The Sources of Invention*, 2nd ed., Macmillan, London.

Johnston, J. (1963). *Econometric Methods*, McGraw-Hill, New York.

Kamien, M. I. and Schwartz, N. L. (1972a). Market structure, rivals' response and the firm's rate of product improvement. *Journal of Industrial Economics*, 20, 159–72.

Kamien, M. I. and Schwartz, N. L. (1972b). Timing of innovations under rivalry. *Econometrica*, 40, 43–60.

Kamien, M. I. and Schwartz, N. L. (1974). Patent life and R & D rivalry. *American Economic Review*, 64, 183–7.

Kamien, M. I. and Schwartz, N. L. (1975). Market structure and innovation: a survey. *Journal of Economic Literature*, 13, 1–37.

Kaplan, N. (1959). The role of the research administrator. *Administrative Science Quarterly*, 4, 20–42.

Kast, F. E. and Rosenzweig, J. E. (1974). *Organisation and Management: A Systems Approach*, 2nd ed., McGraw-Hill Kogakushu Ltd., Tokyo.

Katz, D. and Kahn, R. L. (1966). *The Social Psychology of Organisations*, Wiley, New York.

Kennedy, C. (1964). Induced bias in innovation and the theory of distribution. *Economic Journal*, 74, 541–7.

Klein, B. H. (1962). The decision-making problem in development. In: National Bureau of Economic Research (1962) pp. 477–97.

Klein, B. H. (1965). Policy issues involved in the conduct of military development programs. In: R. A. Tybout (ed.), *Economics of Research and Development*, Ohio State University Press, Columbus, pp. 309–26.

Klein, B. and Meckling, W. (1958). Application of operations research to development decisions. *Operational Research*, 6, 352–63.

Knight, F. H. (1921). *Risk, Uncertainty and Profit*, Houghton Mifflin, Boston.

Knight, K. E. (1967). A descriptive model of the innovation process. *Journal of Business*, 39, 478–96.

Koopmans, T. C. (1949). A reply. *Review of Economics and Statistics*, 31, 86–91.

Kuhn, T. S. (1970). *The Structure of Scientific Knowledge*, 2nd ed., University of Chicago Press, Chicago.

Kuznets, S. (1962). Inventive activity: problems of definition and measurement. In: National Bureau of Economic Research (1962) pp. 19–43.

Langrish, J., Gibbons, M., Evans, W. G. and Jevons, F. R. (1972). *Wealth from Knowledge*, Macmillan, London.

Layton, C. (1972). *Ten Innovations*, Allen & Unwin, London.

Lindblom, C. E. (1959). The science of muddling through. *Public Administration Review*, 19, 79–88.

Little, A. D., Inc. (1963). *Patterns and Problems of Technical Innovation in American Industry*, U.S. Dept. of Commerce, Washington.

Loasby, B. J. (1967). Long range formal planning in perspective. *Journal of Management Studies*, 4, 300–8.

Machlup, F. (1946). Marginal Analysis and Empirical Research. *American Economic Review*, 36, 519–54.

Machlup, F. (1962a). *The Production and Distribution of Knowledge in the United States*, Princeton University Press, Princeton.

Machlup, F. (1962b). The supply of inventors and inventions. In: National Bureau of Economic Research (1962) pp. 143–69.

McKie, J. W. (1972). Industrial organisation: boxing the compass. In Fuchs (1972) pp. 1–15.

MacLaurin, W. R. (1953). The sequence from invention to innovation and its relation to economic growth. *Quarterly Journal of Economics*, 67, 97–111.

Mansfield, E. (1961). Technical change and the rate of imitation. *Econometrica*, 29, 741–66.

Mansfield, E. (1964). Industrial research and development expenditures: determinants, prospects and relation to size of firm and inventive output. *Journal of Political Economy*, 72, 319–40.

Mansfield, E. (1966). National science policy: issues and problems. *American Economic Review, Papers and Proceedings*, 56, 476–88.

Mansfield, E. (1968a). *The Economics of Technological Change*, Norton, London.

Mansfield, E. (1968b). *Industrial Research and Technological Innovation*, Norton, London.

Mansfield, E. (1971). *Research and Innovation in the Modern Corporation*, Macmillan, London.

March, J. G. and Simon, H. A. (1958). *Organisations*, Wiley, New York.

Markham, J. W. (1965). Economic analysis and the research and development decision. In: R. A. Tybout (ed.), *Economics of Research and Development*, Ohio State University Press, Columbus, pp. 67–82.

Marquis, D. G. and Allen T. J. (1966). Communication patterns in applied technology. *American Psychologist*, 21, 1052–60.

Marris, R. (1963). A Model of the managerial enterprise. *Quarterly Journal of Economics*, 77, 185–209.

Marris, R. (1964). *The Economic Theory of Managerial Capitalism*, Macmillan, London.

Marris, R. (1971). The modern corporation and economic theory. In: R. Marris and A. Wood (eds.), *The Corporate Economy: Growth, Competition and Innovative Power*, Macmillan, London, pp. 270–317.

Marschak, J. (1968). Economics of inquiring, communicating, deciding. *American Economic Review*, 58, 1–18.

Marschak, T., Glennan, Jr., T. K. and Summers, R. (1967). *Strategy for R & D: Studies in the Microeconomics of Development*, Springer-Verlag, New York.

Marshall, A. W. and Meckling, W. H. (1962). Predictability of the costs, time and success of development. In: National Bureau of Economic Research (1962) pp. 461–75.

Martino, J. P. (ed.) (1972). *An Introduction to Technological Forecasting*, Gordon & Breach, London.

Medawar, P. (1974). A geometric model of reduction and emergence. In: Ayala and Dobzhansky (1974) pp. 57–63.

Menges, G. (1968). Discussion of the paper by Dr. Winsten. In: K. Borch and J. Mossin (eds.), *Risk and Uncertainty*, Macmillan, London, pp. 136–8.

Minasian, J. R. (1962). The economics of research and development. In: National Bureau of Economic Research (1962) pp. 93–141.

Morton, J. A. (1967). A model of the innovative process (as viewed from a science-based integrated industry). In: *Proceedings of Conference on Technology Transfer and Innovation*, National Science Foundation, Washington, pp. 21–31.

Mueller, D. (1966). Patents, research and development and the measurement of inventive activity. *Journal of Industrial Economics*, 15, 26–37.

Mueller, D. C. (1967). The firm decision process: an econometric investigation. *Quarterly Journal of Economics*, 81, 58–87.

Mulkay, M. J. (1972). *The Social Process of Innovation: A Study in the Sociology of Science*, Macmillan, London.

Näslund, B. and Sellstedt, B. (1972). *Optimal marketing policy and life of products*. Working Paper 72–4, European Institute for Advanced Studies in Management.

Näslund, B. and Sellstedt, B. (1974). Budgets for research and development: an empirical study of 69 Swedish firms. *R & D Management*, 4, 67–73.

National Bureau of Economic Research (1962). *The Rate and Direction of Inventive Activity: Economic and Social Factors*, Princeton University Press, Princeton.

National Science Foundation (1956). *Science and Engineering in American Industry*, Washington.

National Science Foundation (1964). *Decision-Making on Research and Development in the Business Firm*, N.S.F. Reviews of Data on Research and Development, No. 44.

National Science Foundation (1966). *Basic Research, Applied Research and Development in Industry 1963*, N.S.F. 66–15, Washington.

National Science Foundation (1972a). *National Patterns of R & D Resources Funds and Manpower in the United States*, N.S.F. 72–300, Washington.

National Science Foundation (1972b). *Research and Development in Industry 1970*, Surveys of Science Resources Series, Washington.

National Science Foundation (1973a). *National Patterns of R & D Resources Funds and Manpower in the United States 1953–1973*, Washington.

National Science Foundation (1973b). *Research and Development in Industry 1971*, Surveys of Science Resources Series, Washington.

National Science Foundation (1973c). *Science Resource Studies Highlights*, Washington.

Needham, D. (1975). Market structure and firms' R & D behaviour. *Journal of Industrial Economics*, 23, 241–55.

Nelson, R. R. (1959a). The simple economics of basic scientific research. *Journal of Political Economy*, 67, 297–306.

Nelson, R. R. (1959b). Uncertainty, learning and the economics of parallel R & D projects. *Review of Economics and Statistics*, 43, 351–64.

Nelson, R. R. (1962). The link between science and invention: the case of the transistor. In: National Bureau of Economic Research (1962) pp. 549–83.

Nelson, R. R. (1972). Issues and suggestions for the study of industrial organisation in a regime of rapid technical change. In Fuchs (1972) pp. 34–58.

Nelson, R. R. and Winter, S. G. (1974). Neoclassical vs. evolutionary theories of economic growth: critique and prospectus. *Economic Journal*, 84, 886–905.

Nelson, R. R. Peck, M. J. and Kalachek, E. D. (1967). *Technology, Economic Growth and Public Policy*, Brookings Institution, Washington.

Norris, K. P. (1971). The accuracy of project cost and duration estimates in industrial R & D. *R & D Management*, 2, 25–36.

Norris, K. and Vaizey, J. (1973). *The Economics of Research and Technology*, Allen & Unwin, London.

Olin, J. (1973). R & D management practices: chemical industry in Europe. *R & D Management*, 3, 125–35.

Organisation for Economic Co-operation and Development (1970). Excerpt from *The Measurement of Scientific and Technical Activities*, O.E.C.D., D.A.S./S.P.R./70.40 (mimeo), chap. 2, pp. 8–21; reprinted in Freeman (1974) pp. 313–25.

Organisation for Economic Co-operation and Development (1971). *The Conditions for Success in Technological Innovation*, Paris.

Orlans, M. (1973). 'D & R' allocations in the United States. *Science Studies*, 3, 119–59.

Papandreou, A. (1952). Some basic problems in the theory of the firm. In: B. F. Haley (ed.), *A Survey of Contemporary Economics*, Irwin, Homewood, I., vol. II, pp. 183–219.

Peck, M. and Scherer, F. M. (1962). *The Weapons Acquisition Process*, Harvard University Press, Cambridge, Mass.

Penrose, E. (1952). Biological analogies in the theory of the firm. *American Economic Review*. 42, 804–19.

Penrose, E. T. (1959). *The Theory of the Growth of the Firm*, Basil Blackwell, Oxford.

Philips, L. (1971). *Effects of Industrial Concentration: A Cross-Section Analysis for the Common Market*, North-Holland, Amsterdam.

Phillips, A. (1966). Patents, potential competition and technical progress. *American Economic Association Papers*, 56, 301–10.

Prais, S. J. and Houthakker, H. S. (1955). *The Analysis of Family Budgets*, Cambridge University Press, London.

Price, D. J. de S. (1964). The science of science. In: M. Goldsmith and A. MacKay (eds.)., *The Science of Science*, Souvenir Press, London, pp. 195–208.

Price, D. J. de S. (1965). Is technology historically independent of science? A study in statistical historiography. *Technology and Culture*, 6, 553–68.

Price, D. J. de S. (1969). Science and technology: distinctions and interrelationships. In: B. Barnes (ed.), *Sociology of Science*, Penguin Books, London, pp. 166–86.

Price, D. J. de S. (1975). *Science since Babylon*, 2nd ed., Yale University Press, London.

Price, W. J. and Bass, L. W. (1969). Scientific research and the innovative process. *Science*, 164, 802–6.

Quinn, J. B. (1959). Budgeting for research. In: C. Heyel, *Handbook of Industrial Research Management*, Reinhold, New York, pp. 281–313.

Rankin, A. C. (1956). The administrative processes of contract and grant research. *Administrative Science Quarterly*, 1, 275–94.

Rapoport, A. (1974). *Conflict in Man-Made Environment*, Penguin Books, London.

Rapoport, A. and Horvath, W. J. (1959). Thoughts on organisation theory. In: W. Buckley (ed.), *Modern Systems Research for the Behavioural Scientist*, Aldine Publishing Co., Chicago, pp. 71–5.

Rath, G. J. (1967). People, models and R & D organisations. In: Glatt and Shelley (1967) pp. 103–37.

Reeves, E. D. (1958). Development of research budget. *Research Management*, 1, 133–42.

Rosenburg, N. (1974). Science, invention and economic growth. *Economic Journal*, 84, 90–108.

Salomon, J. J. (1971). *The Research System*, vol. I, O.E.C.D., Paris.

Salter, W. E. G. (1960). *Productivity and Technical Change*, Dept. of Applied Economics, Cambridge University.

Sanders, B. S. (1962). Some difficulties in measuring inventive activity. In: National Bureau of Economic Research (1962) pp. 53–83.

Sanderson, M. (1972). Research and the firm in British industry 1919–39. *Science Studies*, 2, 107–51.

Scherer, R. M. (1965). Firm size, market structure, opportunity and the output of patented inventions. *American Economic Review*, 55, 1096–1125.

Scherer, R. M. (1967). Market structure and the employment of scientists and engineers. *American Economic Review*, 57, 524–31.

Scherer, R. M. (1970). *Industrial Market Structure and Economic Performance*, Rand-McNally, Chicago.

Schmookler, J. (1954). The level of inventive activity. *Review of Economics and Statistics*, 36, 183–90.

Schmookler, J. (1962a). Comment on paper by Kuznets. In: National Bureau of Economic Research (1962) pp. 43–51.

Schmookler, J. (1962b). Changes in industry and in the state of knowledge as determinants of industrial invention. In: National

Bureau of Economic Research (1962) pp. 195–232.

Schmookler, J. (1966). *Invention and Economic Growth*, Harvard University Press, Cambridge, Mass.

Schmookler, J. and Brownlee, O. (1962). Determinants of inventive activity. *American Economic Association, Papers and Proceedings*, 52, 165–76.

Schon, D. A. (1967). *Technology and Change: The New Heraclitus*, Pergamon Press, Oxford.

Schon, D. (1971). *Beyond the Stable State: Public and Private Learning in a Changing Society*, Penguin Books, London.

Schott, B. and von Grebner K. (1974). R & D, innovation and microeconomic growth: a case study. *Research Policy*, 2, 380–403.

Schumpeter, J. A. (1943). *Capitalism, Socialism and Democracy*, 3rd edn, Hayer, New York.

Schumpeter, J. A. (1954). *Capitalism, Socialism and Democracy*, 4th ed., Allen & Unwin, London.

Science Policy Research Unit (1972). *Success and Failure in Industrial Innovation*, Centre for the Study of Industrial Innovation, London.

Seeber, N. E., *Research and development decision making in the business firm*. Unpublished manuscript, Graduate School of Industrial Administration, Carnegie Institute of Technology.

Shackle, G. L. S. (1970). *Expectation, Enterprice and Profit: The Theory of the Firm*, Allen & Unwin, London.

Shanks, M. (1967). *The Innovators: The Economics of Technology*, Penguin Books, London.

Sheehan, R. (1966). The way they think at T.R.W. *Fortune*, 74, 153–7, 194–6. Reprinted in: H. I. Ansoff (ed.), *Business Strategy*, Penguin Books, London, 1969, pp. 61–72.

Sherwin, L. W. and Isenson, R. S. (1967). Project Hindsight. *Science*, 156, 1571–7.

Shubik, M. (1970). A curmudgeon's guide to microeconomics. *Journal of Economic Literature*, 8, 405–34.

Siegel, I. H. (1962). Scientific discovery and the rate of invention. In: National Bureau of Economic Research (1962) pp. 441–57.

Simon, H. A. (1955). A behavioural model of rational choice. *Quarterly Journal of Economics*, 69, 99–118.

Simon, H. A. (1959). Theories of decision-making in economics and behavioural science. *American Economic Review*, 49, 253–83.

Simon, H. A. (1965). *The Shape of Automation for Men and Management*, Harper & Row, New York.

Simon, H. A. (1969). *The Sciences of the Artificial*, M.I.T. Press, Cambridge, Mass.

Smith, W. J. J. and Creamer, D. (1968). *R & D and Small Company Growth: A Statistical Review and Company Case Studies*, National Industrial Conference Board Inc., New York.

Smyth, D. J., Samuels, J. M. and Tzoamnos, J. (1972). Patents, profitability, liquidity and firm size. *Applied Economics*, 41, 77–86.

Staff Report (I.R.I) (1962). Proceedings of Industrial Research Institute Study Group Meetings, No. 6: Expediting the Transition from Research to Production. *Research Management*, 5, 33–47.

Starbuck, W. H. (1965). Organisational growth and development. In: Starbuck (1971) pp. 11–141, Penguin, London.

Taplin, W. (1959). Advertising appropriation policy. *Economica*, 26, 227–39.

Taton, R. (1957). *Reason and Chance in Scientific Discovery*, Hutchinson, London.

Terreberry, S. (1968). The evolution of organisational environments. *Administrative Science Quarterly*, 12, 590–613.

Thomas, C. A. (1963). Research and development: a challenge for management. *Research Management*, 6, 305–11.

Thorpe, W. H. (1974). Reductionism in biology. In: Ayala and Dobzhansky (1974) pp. 109–36.

Tilles, S. (1969). Making strategy explicit. In: H. I. Ansoff (ed.), *Business Strategy*, Penguin Books, London, 1969, pp. 180–209.

Trattner, E. and Zidaroiu, C. (1972). Economic optimization of technical-scientific research plans. *R & D Management*, 2, 83–6.

UNESCO (1970). Excerpts from measurement of output of research and experimental development, chaps. 2–6 and tables, pp. 10–43; reprinted in Freeman (1974) pp. 332–87.

Vining, R. (1949a). Koopmans on the choice of variables to be studied and of methods of measurement. *Review of Economics and Statistics*, 31, 77–86.

Vining, R. (1949b). A rejoinder. *Review of Economics and Statistics*, 31, 91–4.

Weiss, P. A. (1969). The living system: determinism stratified. In: A. Koestler and J. R. Smythies (eds.), *Beyond Reductionism: New Perspectives in the Life Sciences*, Hutchinson, London, pp. 3–55.

Williams, B. (1961) 'Comment'. In: Federation of British Industry, *Industrial Research in Manufacturing Industry 1959–60*, London, pp. 21–8.

Williamson, O. E. (1964). *The Economics of Discretionary Behaviour:*

Managerial Objectives in a Theory of the Firm, Prentice-Hall, Eanglewood Chiffs, N. J.

Williamson, O. E. (1970). *Corporate Control and Business Behaviour*, Prentice-Hall, Englewood Cliffs, N. J.

Williamson, O. E. (1971). Managerial discretion, organisation form, and the multi-division hypothesis. In: R. Marris and A. Wood (eds.), *The Corporate Economy: Growth, Competition and Innovative Potential*, Macmillan, London, pp. 343–86.

Wilson, A. (1973). *The Concorde Fiasco*, Penguin Books, London.

Winsten, C. B. (1968). Uncertainty and the communication of information. In: K. Borch and J. Mossin (eds.), *Risk and Uncertainty*, Macmillan, London, pp. 123–34.

Winter, S. G. (1964). Economic 'natural selection' and the theory of the firm. *Yale Economic Essays*, 4, 225–77.

Winter, S. (1971). Satisficing, selection and the innovating remnant. *Quarterly Journal of Economics*, 85, 237–61.

Worley, J. S. (1961). Industrial research and the new competition. *Journal of Political Economy*, 69, 183–6.

Wu, S. Y. and Pontney, J. A. (1967). *An Introduction to Modern Demand Theory*, Random House, New York.

Index